DICTIONARY OF STATISTICS

DICTIONARY OF STATISTICS

S.H. Banday
M.A. Shah

ANMOL PUBLICATIONS PVT. LTD.
NEW DELHI - 110 002 (INDIA)

ANMOL PUBLICATIONS PVT. LTD.
4374/4B, Ansari Road, Daryaganj
New Delhi - 110 002
Ph.: 23261597, 23278000
Visit us at: www.anmolpublications.com

Dictionary of Statistics

First Edition, 1989

Reprint, 1990, 1992, 1994, 1996, 1998, 2002

Reprint, 2005

PRINTED IN INDIA

Published by J.L. Kumar for Anmol Publications Pvt. Ltd., New Delhi - 110 002 and Printed at Mehra Offset Press, Delhi.

Preface

This dictionary includes the most frequently used terms which have been carefully selected from the various fields like management, commerce, mathematics, economics, econometric, business and industry, sociology, science and engineering.

Every effort has been carefully made to write the entries in a clear and lucid style to provide both straight forward definitions and invaluable background information. At some appropriate places, line diagrams have been included whenever the meaning of a word can be best understood by means of a diagram.

This dictionary will be of immense value to undergraduate and post-graduate students and to those appearing for competitive examinations, research scholars, teachers administrators and policy makers.

When a dictionary of this kind is being compiled, it becomes essential to draw upon the work of many authorities and seek the advice of colleaques to all of whom the editors are deeply indebted.

Finally, the editors express their sincere thanks to the publishers and printer for printing this book promptly.

All comments from users on omissions or shortcomings will be most welcome.

EDITORS

A

Abnormal Curve (Curva Anormale). Refers to a frequency curve which is having abnormality, but differing from a Normal curve with the same median and dispersion about the median. If the frequency curve is able to cross this corresponding Normal curve more than once in at least one of the two branches on either side of the median, it is regarded to be complex abnormal (*anormale complessa*). If it is symmetrical and in each branch crosses the Normal curve only once it is regarded simple abnormal (*anormale semplice*).

Abnormality (Abnormalita). Refers to a measure of dissimilarity from the Normal curve, which is somewhat similar to a measure of kurtosis.

Abrupt Distribution. A continuous frequency distribution has been considered to be abrupt at a finite terminal point of its range when the frequency and the first derivative of the frequency have not been both zero at that point.

Abscissa. The horizontal or x coordinate in a two-dimensional system of Cartesian coordinates. The vertical or y-coordinate is termed as the *ordinate*. See Fig. 1.

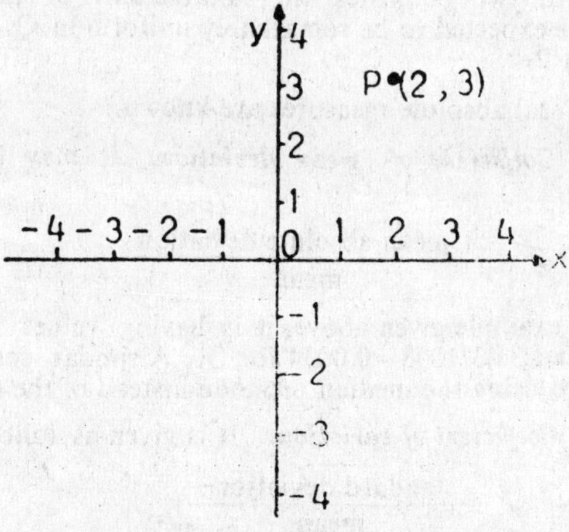

Abscissa. In this example, the abscissa of P has been 2, and the ordinate of P has been 3.

Absolute Deviation. Refers to the difference which is taken without regard to sign, of a variate value from some given value which may be a constant, a parent value or a sample value.

Absolute Difference. Refers to the difference which is taken without regard to sign, between the values of two variables ; and in particular of two random variables.

Absolute Error. The absolute error of an observation x has been the absolute deviation of x from its 'true' value.

Absolute Frequency. Refers to the actual frequency of a variate, as distinct from the relative frequency, namely the ratio of the frequency to the total frequency of all variate values.

Absolute Measure. Refers to a measure of variate values which is independent of origin and scale.

Absolute Measure of Dispersion. Refers to a measure of dispersion (or spread) with reference to the size and distribution of the figures involved.

The sets of numbers P and Q are given below :

P	2	4	6	8	10	12	14
Q	1002	1004	1006	1008	1010	1012	1014

Both are having the same range (12), interquartile range (8) and standard deviation (4). Relative to the sizes of the numbers, however, P has been clearly much more spread than Q. If, for instance, the figures represented typical incomes of people in two countries (in suitable units of currency), they might be expected to be remarkably uniform in Q but not at all so in P.

Several absolute measures are known :

(*a*) *Coefficient of mean deviation.* It may be given as follows :

$$\frac{\text{mean absolute deviation}}{\text{mean}}$$

For the example given above, it is having values 3.43/8=0.43 for P, and 3.43/1008=0.0034 for Q. A similar coefficient can be found using the median or mode instead of the mean.

(*b*) *Coefficient of variation.* It is given as follows :

$$\frac{\text{standard deviation}}{\text{mean}}$$

For P, it is having the value 4/8=0.5 ; for Q, 4/1008=0.00397. Coefficient of variation has been sometimes written as a percentage.

$$\frac{\text{standard deviation}}{\text{mean}} \times 100\%$$

For P, it is having value 50% ; for Q, 0,397%.

(c) *Quartile coefficient of dispersion.* It is given as follows :

$$\frac{\text{upper quartile}-\text{lower quartile}}{\text{upper quartile}+\text{lower quartile}}$$

For P, the quartiles have been 12 and 4 ; for Q, 1012 and 1004. Hence the values of the quartile coefficient of dispersion would be as follows :

for P, $\frac{12-4}{12+4}=0.5$

for Q, $\frac{1012-1004}{1012+1004}=0.00397$

Absolute Number. Refers to any number as such, simply as a number, without reference to any particular objects whatever except in so far as these objects possess the number property.

Absolute Value of a Number. The value of a number without regard to positive or negative (plus or minus) sign. The absolute value of both +5 and −5 has been 5.

Absolute Value or Modulus, *n*. Refers to the magnitude of a real number or quantity, regardless of its sign. Symbol : straight brackets | |.

$|-7|=7$ $|18.3|=18.3$

Absorbing State. The term used for a state in a stochastic process which, once reached, cannot get quitted. It is also called an absorbing barrier, but this term is generally used where the process terminates when the barrier gets reached.

Absorption Distributions. Refers to the distributions that represent the number of 'individuals' (*e.g.*, particles) that are unable to cross a specified absorbing region.

Acceptable Quality Level (AQL). The term used for the proportion of effective units in a batch which is considered as desirable by the consumer of the batch ; the complement of the proportion of defectives which he wishes to tolerate.

Acceptable Reliability Level. Refers to a concept which is analogous to acceptable quality level, in which the characteristic has been expressed in terms of failure rate per unit time, *e.g.*, 1000 hours.

Acceptance Control Chart. It is a simple form of quality control chart in which lots have been accepted or rejected in accordance to whether plotted values of the mean value of the lot or batch fall inside or outside the specified acceptance control limits.

Acceptance Error. Refers to an alternative name for β error.

Acceptance Line. In sequential analysis it refers to the graph of the acceptance number as ordinate against the sample number as abscissa. It is also termed as the acceptance boundary. There has been a corresponding rejection line.

Acceptance Number or Allowable Defects. Refer to the greatest number of defectives in a sample that will permit the batch (from which the sample is drawn) to be accepted without further quality control testing. Symbol : **n.**

Acceptance Region. In the theory of testing hypotheses, it refers to a region in the sample space such that, if a sample point falls within it, the hypothesis under test has been accepted.

Accumulated Deviation. The term used for the graduation of a grouped frequency distribution provides, for each grouping interval, a 'theoretical' or 'expected' frequency. The difference of this quantity and the observed quantity taken with regard to sign and cumulated for increasing values of the variates starting from the least, is termed as the accumulated deviation. It has been equivalent to the difference between the observed and theoretical distribution functions of the grouped distributions.

Accumulated Process. Refers to a stochastic process which has been derived from the cumulated sum of random variables. For example, the total damage during time t of a sequence of thunderstorms is forming an accumulated process.

Accumulation Point. An accumulation point of a set of points is a point P such that there is at least one point of the set distinct from P in any neighbourhood of the given point ; a point which is the limit of a sequence of points of the set (for spaces which satisfy the first axiom of countability). An accumulation point of a sequence is a point P such that, for any integer n, each neighbourhood of P contains at least one term of the sequence after the nth term ; *e.g.*, the sequence 1, $\frac{1}{2}$, 1, $\frac{1}{3}$,

1, $\frac{1}{4}$, 1, 1/5, ... has two accumulation points, the numbers 0 and 1 (also see sequence—accumulation point of a sequence). *Syn.* cluster point, limit point. See Bolzano-Bolzano Weierstrass theorem, condensation—condensation point.

Accuracy. In the general statistical sense this term denotes the closeness of computations or estimates to the exact or true values. In a more specialised sense the word is also having meaning : (*a*) in relation to an estimator, *unbiassedness* (see unbiassed estimator) ; (*b*) in relation to the reciprocal of the standard error, the precision. Neither usage can be recommended.

Adaptive Inference. When we collect data over a period of time, the character of the inferences made at any one point will depend on the nature of previous observations and is then said to be adaptive. In a more restricted sense, the observations from a single sample may themselves determine the type of statistic used in obtaining inferences about the parent population.

Adaptive Optimisation. The term used for a system of control for physical processes. It is based upon statistical adaptation as a series of observations proceeds through time. It is having a particular application in the field of statistical quality control.

Addition Law. It is a law which is concerning the probabilities of two events occurring. The addition law states that :

$$p(\text{A or B}) = p(\text{A}) + p(\text{B}) - p(\text{A and B})$$

It is usually written, using set notation, as :

$$p(A \cup B) = p(A) + p(B) - p(A \cap B)$$

Addition Theorem. The principle holding that the probability that either or any one of two or more mutually exclusive events will occur in a given trial or occurrence is the sum of the individual probabilities.

Additive Model. It is a model in which the factors influencing the dependent variable are having an additive effect.

Additive (Random Walk) Process). It is a stochastic process having independent increments, that is to say, a process $[x_t]$ has been additive if, for $t_1 < t_2 < \ldots < t_n$, the differences $x_{t_2} - x_{t_1}$, $x_{t_3} - x_{t_2}$, etc., have been independent.

Admissible Decision Function. In the general theory of statistical decision functions an admissible decision function may be defined as one for which it can be shown that there has been no decision function which has been uniformly better.

Admissible Hypothesis. Generally, a hypothesis has been considered to be admissible when it has been possible within the conditions of the problem. More specifically, if there has been a distribution of known mathematical form depending on k unknown parameters θ then a statistical hypothesis could be stated by specifying any k set of θ's which has been *a priori* possible. Such hypothesis form the set of admissible hypotheses.

Admissible Test. A test of a hypothesis is considered to be admissible if, in respect of a particular class of alternatives, there has been no other test with uniformly greater power ; *e.g.*, if a uniformly most powerful test exists, no other test will be admissible.

Affinity. A concept which refers to a measure of divergence or distance between two statistical populations. It gives, for example, a generalisation of Mahalanobis distance.

Age. 1. Refers to the period of time that a person, animal or plant has lived, or that an object has existed, *i.e*, the *chronological age*. 2. Also, refers to the level in year that a person has reached in any area of development compared with the normal level for his chronological age, for example, *mental age*, *reading age*.

Age-sex Pyramid. A chart showing in one diagram the age distribution of males and females in a particular population. It is actually two histograms, placed back to back, with the bars lying horizontally, rather than standing vertically. A vertical line divides the chart into two parts. The left side of the chart, with its bars extending to the left from the vertical line, usually indicates the number or percentage of males in the particular age category represented by each bar. The right side, with bars extending to the right from the mid-line, provides the same information for females. This type of chart is sometimes called an age-and-sex triangle. See Histogram ; population pyramid.

Age Specific Death Rate. A term which is generally used in life analysis of human or biological populations for denoting the probobility of dying in the next unit time period of (an entity that had survived for a stated period of) time. Substitution of word 'failure' for 'death' makes the term useful in connection with physical systems or components.

Aggregate. Refers to the value of a single variable which is resultes from the combination of data for a number of variables. For example, an examination mark has been often an aggregate of the marks of several different papers, and perhaps a project. The process of forming an aggregate has been termed as *aggregation.*

Aggregation. A word which is used to denote the compounding of primary data into an aggregate, generally for the purpose of expressing them in a summary form.

Aggregative Index. Refers to an index-number which has been constructed by aggregating a number of items (as distinct, for example, from picking out a representative item). In price indices, if p_o typifies the prices in the base period and p_n those for the current period and q_o and q_n typify the quantities in the base and current periods respectively, then the two principal forms of aggregative index will be as follows :

$$\frac{\Sigma\,(p_n q_o)}{\Sigma\,(p_o q_o)} \text{ and } \frac{\Sigma\,(p_n q_n)}{\Sigma\,(p_o q_n)}$$

where the summation occurs over the commodities, to be aggregated. [Also, refer to Laspeyres' index, Paasche's index.]

Alias. This term finds use in two branches of statistical analysis. In connection with design of experiments, when a factorial design has been only fractionally replicated certain comparisons are not able to distinguish between some of the treatment combinations ; these are considered to be aliases.

In connection with the spectrum analysis of time series, an analogous concept arises since a single width of the interval of observation does not allow a distinction between certain angular frequencies.

Alienation. Refers to the effect of sampling errors on a correlation, coefficient.

Allocation, of a Sample. Refers to the way in which sample numbers have been assigned to various parts of a population by the sampling plan ; *e.g.*, for a stratified population it could be decided to allocate the total sample number to the strata in proportion to the numbers of individuals in those strata.

Allowable Defects. In quality control, the number of allowable defects in a sample may be equal to the critical number (designated by the particular scheme of sampling inspection) such that, if this number gets exceeded, the whole of the remainder of the inspection-lot should be examined or the lot rejected out of hand. [See also Acceptance Number.]

Almost Certain. A probability dependent on a parameter n may tend to unity as n tends to infinity. The event to which it relates is then considered to be 'almost certain' in the limit. More generally, if the measure of an event has been unity and its converse has measure zero but does not get related to an empty set, the event has been 'almost' certain.

Almost Certainly. See almost everywhere.

Almost Everywhere. A statement about a value x in a domain is considered to be true almost everywhere with respect to a probability measure μ on x if the set of values of x for which the statement has been false has been of measure zero. Alternatively the statement is considered to be true almost surely, or almost certainly.

Almost Stationary. A term which is used to cover the situation where the series resulting from wide interval sampling of a non-stationary time series has been itself stationary.

Alpha Error or α-Error. In the theory of testing hypotheses it is an error incurred by rejecting an hypothesis when it has been actually true : more usually known as error of the first kind or Type I Error. In Quality Control it has been equivalent to producer's risk.

Alternating Renewal Process. Refers to a special case of the two-state semi-Markov process in which the two types of intervals are appearing alternately.

Alternative Hypothesis. The conclusion we accept when the data fail to support the null hypothesis.

Amount of Inspection. In quality control, is refers to the number of items (size of sample) taken from each lot and inspected in accordance to the particular sampling plan being used.

Analysis of Dispersion. It is generalisation of the univariate analysis of variance providing a theory for multivariate tests of significance. It is sometimes termed as multivariate analysis of variance.

Analysis of Variance or Anova, *n*. It is a technique for investigating how much of the variability in a set of observations could be ascribed to different causes. In a ***one-way analysis of variance***, many different samples are drawn. The data would be then tested to see if its variability (as measured by its variance) is all random or if parts of it is the result of systematic differences between the samples (see factorial experiment). Thus one-way analysis of variance tests the null hypothesis that all the samples have been drawn from the same parent population. (In a *two-way analysis of variance* the method has been extended to variation for two factors.)

Analysis of variance has been has been involving separating the sample variance into two components, that within the samples and that between them. These are then compared using the F-test. The variance is separated using the sum of squares result.

$$\sum x_T^2 = \sum x_B^2 + \sum x_W^2$$

Total sum of squares=That between the groups+That within them.

Analysis of variance may be carried out on the ranks of the data, rather than the actual values.

Analysis of Variance for Regression. The procedure for computing the F-ratio used to test the significance of the regression as a whole.

Analytic Regression. Refers to a regression relation where the independent variables (regressors) are formed by polynomials, trigonometric sums or other analytic expressions.

Analytic Survey. It is a (sample) survey where the primary purpose of the design has been the comparison between sectors or subgroups of the population sampled.

Antilogarithm *n.* **Antilogarithm of a Given Number.** The number whose logarithm is the given number ; *e.g.,* $\text{antilog}_{10} 2 = 100$. *Syn.* inverse logarithm. To find an antilogarithm corresponding to a given logarithm that is not in the tables, subtract the next smaller mantissa from the given one and from the next larger one, divide the former difference by the latter and annex the quotient to the number corresponding to the smaller mantissa. See Logarithm.

Antimode. Refess to the variate value, if any, for which a frequency distribution is having a minimum. The expression is generally confined to the case where the minimum, is not zero and is a true minimum, *e.g.* the zero tails of a Normal distribution at infinity are not antimodes in the sense.

AOQ. Average outgoing quality.

AOQL. Average outgoing quality limit.

Approximation. Refers to the process or result of making a rough calculation, estimate or guess. Approximation is used extensively when deciding if an answer has been reasonable. The statement 'It takes five hours to fly from Britain to Austrialia by jumbojet' is evidently false. The distance is about 20,000 km, the speed about 1,000 km per hour, and so the time taken is about 20 hours. To decide that the statement was false, it was not necessary to know exact values of the speed of the aeroplane or the distance involved.

When numerical information is being given, the number of significant figures should be consistent with the accuracy of the information involved. Thus the statement 'The cost of running my car is Rs. 1,783 47 per year' is misleading because it is not possible to estimate it so accurately, especially when depreciation is taken into consideration. A better figure would have been approximately Rs. 1,800', which is obviously a round number, indicating an error of perhaps Rs. 50 either way.

Approximation Error. Refers to an error due to approximation in numerical calculations as distinct, for example, from an error of observation. More particularly, it refers to a rounding error.

A Priori Probability. Probability estimate made prior to receiving new information.

Area Diagram. A type of graphic presentation in which a figure, such as a circle or rectangle, is divided into sections, each proportional in area in the size of the quantity it represents. The pie chart is a common type of area diagram.

Area Sampling. A method of sampling used when no complete frame of reference is available. The total area under investigation is divided into small sub-areas which are sampled at random or by some restricted random process. Each of the chosen sub-areas is then fully inspected and enumerated, and may from a frame for further sampling if desired. The term may also be used (but is not to be recommended) as meaning the sampling of a domain to determined area, *e.g.*, under a crop.

Arithmatic Distribution. Refers to a discontinuous distribution for which the variate values have been equidistant and can be represented as $a \pm \lambda b, \lambda = 0, 1, 2, \ldots$

In the multivariate analogue the expression 'lattice distribution' has been used.

Arithmetic Mean. Refers to the result which is obtained by adding the numbers or quantities in a set and dividing the total by the number of members in the set. It is often called average. For example, the arithmetic mean of 43, 49, 63, 51 and 28 has been :

$$\frac{43+49+63+51+28}{5}=46.8$$

The arithmetic mean of the members $x_1, x_2, ...x_n$ has been given by :

$$\frac{x_1+x_2...+_n}{n}$$

It may also be written as :

$$\frac{1}{n}\sum_{i=1}^{n} x_i \quad \text{or} \quad \frac{\sum_{i=1}^{k} x_i f_i}{\sum_{i=1}^{k} f_i}$$

where f_i is the frequency of the value x_i, and there are k distinct values of x, See also mean.

In current English usage the word 'arithmetic' it frequently omitted so that where a 'mean' is mentioned the arithmetic mean is to be understood.

Array. In the most general sense it refers to an explicit display of a set of observations. More usually, the term denotes some special arrangement of the observations, *e.g.*, in order of magnitude. A *frequency array* may be defined as an array of frequencies according to variate values, that is to say, a frequency distribution. The term 'array' has been often used for the individual frequency distributions which form the separate rows and columns of a bivariate frequency table.

Artifact Statistical. A statistical inference resulting from bias in the collection or treatment of data.

Ascertainment Error. See non-sampling error.

Association. Refers to the tendency of two events to occur together. If the probabilities (or frequencies) of the different combinations of events A and not A, and B and not B, have been

	A	not A
B	a	b
and B	c	d

the association is considered to be positive if $ad>bc$, and negative if $ad<bc$. If $ad=bc$,, the events A and B are independent. *Yule's coefficient of association* may be given as follows :

$$\frac{ad-bc}{ad+bc}$$

Association, Statistical. Refers to a statisticai relationship in which two or more variables occur or very together. Any of a variety of statistical techniques could be used to measure the extent to which a change in one variable results in a corresponding change in another variable or (variables). The selection of a particular technique is dependent upon such factors as whether the variables are qualitative or quantitative, the number of variables, and the nature of the data.

Assumed Mean. Refers to an estimated or approximate value for the arithmetic mean, or average which is used to simplify its its calculation.

When the arithmatic mean of a set of numbers of similar size is being found out, the calculation can sometimes be simplified by :

(*a*) Assuming a mean.

(*b*) Working out the differences from it (+or−) for each number ;

(*c*) Knowing the mean of the differences ;

(*d*) Adding the mean of the differences into the assumed mean.

It does not matter what value is taken for the assumed mean, but the nearer it is to the true mean, the smaller are the numbers involved.

Example : find the mean of the ages of six children who are, in years and months, 13y11m, 14y4m, 13y6m, 14y6m, 14y1m and 14y2m,

(*a*) Consider an assumed mean : 14y0m :

(*b*) Find out the differences from the assumed mean : −1, +4, −6, +6, +1 and +2 months ;

(*c*) **Find out the mean of the difference :**

$$\frac{-1+4-6+6+1+2}{6}=+1 \text{ month}$$

(*d*) **The mean age of the children would be thus given by :**

true mean = assumed mean + mean of differences

= 14y0+1m

The mean age of the children would be 14y1m.

Assumed Mean. An arbitrary origin or working mean which is used for the calculation of moments. The expression has not been to be recommended.

Asymmetrical Distribution. Refers to a distribution which is not symmetrical, that is to say, for which there is no central value a such that $f(x-a)=f(a-x)$, $f(x)$ being the frequency function.

Asymptotic Distribution. Refers to the limiting form of frequency or probability distribution which is dependent on a parameter, such as sample-number or time, as that parameter tends to limit, usually infinity.

Asymptotic Normality. A distribution dependent on a parameter n, generally a sample number, is considered to be asymptotically Normal if, as n tends to infinity, the distribution tends to the Normal form.

Asymptotic Standard Error. The standard error of a statistic nearly always depend on the sample number n. If, as n tends to infinity, the standard error S is asymptotically equivalent to a quantity Q, *i.e.* is such that S/Q tends to unity, then Q may be termed the asymptotic standard error. As so defined, Q is not unique but in the vast majority of cases S^2 is expressible as a term in n^{-1} plus terms of lower order in n, and the first is taken as the square of Q.

Attenuation. Where observations on bivariate material have been subject to errors of measurement the true correlation between the variates will be obscured, usually being underestimated. The correlation is then considered to be attenuated. In 1904 Spearman advanced the following formula which would correct this attenuation

$$r'_{xy}=\frac{r_{xy}}{\{r_{xx}\, r_{yy}\}^{1/2}}$$

where r_{xy} denotes the geometric mean of correlations between independent determinations of x and y ; r_{xx} and r_{yy} have been the means of correlation between independent determinations of x and of y and r'_{xy} is the corrected correlation. [See reliability coefficient.]

Attenuation, Statistical. The reduction in the size of a correlation through errors of measurement. Because of errors the correlation obtained in a study is lower than the true correlation that exists in the population. A correction for attenuation may be used to estimate the size of the correlation without the chance errors. The obtained correlation is divided by the square root of the product of the reliability coefficients of the two variables. The resulting value is the maximum probable correlation between the variables. It should be remembered, however, that this is a theoretical and not an obtained value. See coefficient of reliability.

Attribute. An inherent property, usually one that is measured by a finite number or a discrete set of values. Sometimes used as the outcome of one of a set of mutually exclusive events (*e.g.*, the sex of person, or the weight to the nearest 10 lb) ; when used in this sense, attribute space is the same as sample space. Often attribute sampling refers to an experiment with only two outcomes, *e.g.*, classifying an object as defective or nondefective.

Attribute Testing. Refer to the testing of a product for quality control when the result has been either 'good' or 'defective'. Attribute testing could be contrasted with variable testing where the result is a quantitative measure, like the length of a nail or the resistance of an electrical component. See also lattice diagram.

Autocorrelation. Refers to the internal correlation between members of series of observations ordered in time or space.

Autocorrelation Function. The autocorrelation function of a stationary stochastic process may be defined as the autocovariance divided by the variance.

Autonomaus Equations. In econometrics, refers to an equation which is describing the behaviour of one particular group or sector of the economy, *e.g.*, a demand equation describing only buyer's behaviour, is regarded to be autonomous if it is affected only by changes in the behaviour of that particular group or sector. Equations of this kind considered sometimes termed 'structural equations' but this is not be, recommended, as such equations may include variates from the whole system.

Autoregression. Refers to the generation of a series of observations whereby the value of each observation has been partly dependent upon the value of those which have immediately preceded it, *i.e.*, each observation stands in a regression relationship with one or more of the immediately preceding terms. A scheme of autoregression may be regarded to be as a stochastic process of a conditional kind.

Autoregressive Model. Refers to an econometric model which is based upon the autoregressive process but also containing lagged versions of some or all of the endogenous variates appearing in the model specification.

Autoregressive Series. Refers to a series which was generated by an autoregressive process. It is the realisation of an autoregressive process.

Autoregressive Transformation. If there has been autocorrelation in the error term of an autoregressive process it has been sometimes possible to transform the original variates to new variates such that the autoregressive scheme in the transformed variates is having an uncorrelated error term. This procedure is termed as an autoregressive transformation.

Average. Refers to a numerical value that is a central point about which other values in a series get dispersed. An average provides a single, summary measure of a distribution of numerical values. In statistical terms it is usually referred to as a measure of central tendency, for it summarizes in one value the central tendency of all the variations in a distribution taken as a whole. There have been different types of averages, or different measures of central tendency, each of which may result in a different average value for the same series of values.

Average Absolute Deviation. In a data set, it refers to the average distance of the observations from the mean. It is given as follows :

$$\text{Average absolute deviation} = \frac{\Sigma \mid x-\mu \mid}{N}$$

Here x represents the item or observation, μ the population mean, N the number of items in the population and $\Sigma(x-\mu)$ the sum of all the values of $(x-\mu)$.

This formula enables us to calculate the average absolute deviation for a population. Because this measure of variability deals with the absolute value of the difference between each item in the data set and the mean it is not as useful for further calculation as is the variance, which equates each distance.

Average Deviation. A synonym of mean deviation, not to be recommended. It is sometimes called the ***mean deviation***, it is a simple measure of dispersion of a series of figures from their arithmetic mean. It is equal to the sum of the deviation from the mean, with sign ignored, divided by the number of figures in the series.

$$\text{Average deviation} = \frac{\Sigma x}{n}$$

where Σ is the sign of summation, x the deviation from the mean, and n the number of figures in the series. Statisticians prefer the *standard* deviation. It is having wide and general application in all fields except statistical theory.

Average of Relatives. An average which is usually in the form of an index number, of a set of *relatives*, that is to say values obtained as the ratio of a magnitude in the given period to the corresponding magnitude in the base period. In price index numbers, the price relatives are generally weighted by the values either of the base period or of the given period. Where the weights used have been the values in the base year the formula reduces to that of Laspeyres. If values in the given period have been tied with a harmonic average the formula reduces to that of Paasche.

Average Outgoing Quality (**AOQ**). Refers to the proportion of defective items being sent out with a product after a quality-control scheme has been implemented. For a single inspection scheme (single sampling scheme) with acceptance number n from samples of size N, with 100% inspection of failed batches AOQ may be put as follows :

$$\text{AOQ} = \text{P}(p)\ (1 - n/\text{N})p$$

where, p refers to the proportion of defectives manufactured, and P(p) refers to the probability of a batch with proportion p defective being accepted.

Average Outgoing Quality Limit (**AOQL**). Refers to the highest average proportion of defective items that will be let through in the long run, following a quality-control sampling scheme.

When we plot average outgoing quality (AOQ) against the proportion of defectives manufactured, p, the graph rises to a maximum and then falls away. A typical graph for a single sampling scheme has been deficted in Fig. If p is small, the AOQ will be low because there are few defectives to get through. If p is large, the sampling scheme is likely to result in most batches being stopped and undergoing 100% inspection ; any defectives are then rejected. Between these two

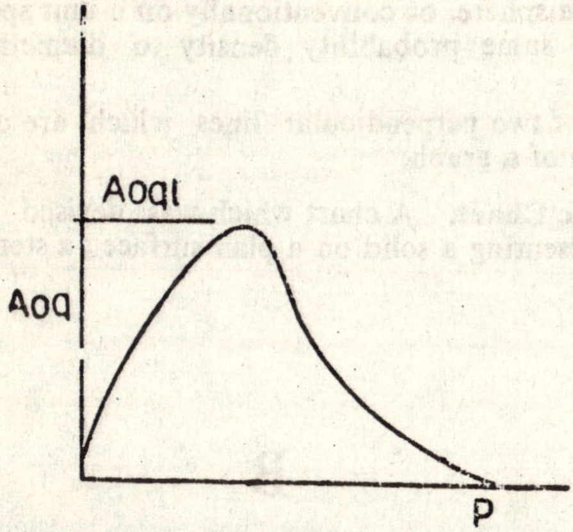

Average outgoing quality limit (AOQ). The average outgoing quality (AOQ) of goods is plotted against the proportion of defective goods (p) to find the maximum value (AOQL).

cases have been values of p which are neither very small, nor large enough to involve most batches being stopped.

The maximum value shown on such a graph is known as the average outgoing quality limit, and represents the highest average proportion of defectives that will be let through in the long run. It has been however, possible for worse individual batches to get through.

Average Run Length (ARL). The average run length of a sampling inspection scheme at a given level of quality may be defined as the average number of samples of n items taken in the period between the time when the process commences to run at the stated level and that at which the scheme reveals a change from acceptable to rejectable quality level is likely to have occurred. This concept must not get confused with average article run length.

Average Sample Number Curve. Refers to the graph of the average sample number functton, with the function as ordinate against the parameter as abscissa.

Average Sample Run Length. An alternative name which is sometimes used for average run length so as to distinguish it from the concept of average article run length.

Axial Distribution. It is a spherical distribution which describes the relative frequencies of the possible orientations in three dimensions of a random axis. As an axis is a direction without a sense, such a distribution can be defined entirely on a unit hemisphere, or conventionally on a unit sphere by assigning the same probability density to diametrically opposite points.

Axis. One of two perpendicular lines which are used in the construction of a graph.

Axonometric Chart. A chart which was devised for the purpose of representing a solid on a plan surface ; a stereogram.

B

Bahadur Efficiency. Refers to an approximate measure of asymptotic relative performance for sequences of test statistice.

Balanced Differences. In a systematic sample of an ordered series the selected units are not located at random and therefore no fully valid estimate of sampling error would be possible in general. One method of overcoming this difficulty has been to construct artificial strata by dividing the series into 'blocks' of equal length and to consider the members falling within a block as having been chosen at random within that block.

If there happen to be only two members in the block an estimate of error will be based on their difference. If there happen to be more it may be based on more complicated linear functions designed to eliminate systematic error, *e.g.*, for seven members there would be used 'balanced differences' of the form.

$$d=\tfrac{1}{2}y_1-y_2+y_3-y_4+y_5-y_6+\tfrac{1}{2}y_7.$$

The numbers of terms has been arbitrary but seven or nine will eliminate most of any systematic component of variation.

Similar ideas have been found to be applicable to systematic sampling in more than one dimension.

Balanced Sample. When the mean value of some characteristic is known for a population and the value of the characteristic can be ascertained for each member of a sample, it becomes possible to choose the sample so that the mean value of the characteristic in it approximates to the parent mean. Such a sample is considered to be balanced.

The object of balancing is to get a sample which is representative of the parent in respect of some other characteristic for which the parental value is not known. Authorities have been reported to differ about the value and validity of the method.

Ballot Theory. Refers to the various generalisations and extensions of a theorem by Bertrand (1887) : when a candidate (A) scores a votes and candidate (B) scores (b) votes, $a > \mu b$ and μ is non-negative integer, then the probability throughout the counting that the votes for (A) are always greater than μ times the votes for (B) will be $Pr = (a - \mu b)/(a + b)$.

Bar Chart or Bar Diagram. A display of data using a number of rectangles, of the same width, whose lengths are proportional to the frequencies they represent. Bar charts can be drawn in several ways ; for example, with the bars drawn vertically or horizontally (Figs. 1, 2) immediately next to each other, or with a fixed gap between them. A *component* or *compound bar chart* has two or more parts to each bar (see Fig. 3). There are many ways in which such diagrams can be drawn, some of them very artistic. The lengths of the bars and the areas of the constituent parts are *pro rata* to the absolute magnitudes of the statistics that they represent.

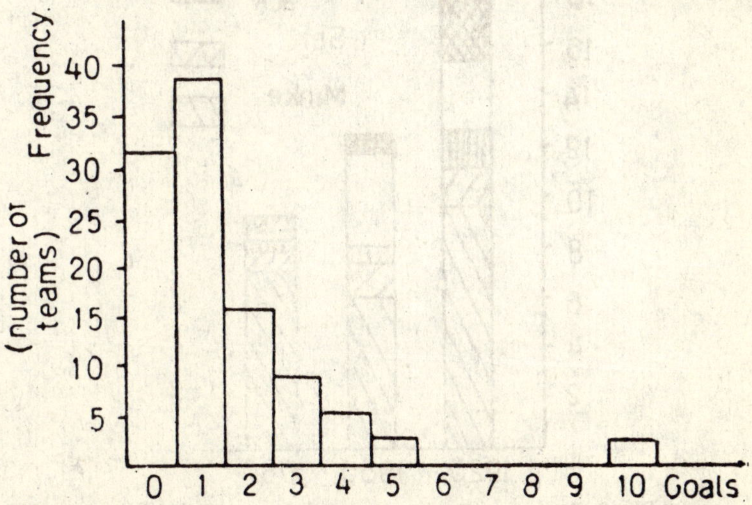

Fig. 1. Bar chart. Goals scored in the World Cup finals in 1982 in Spain

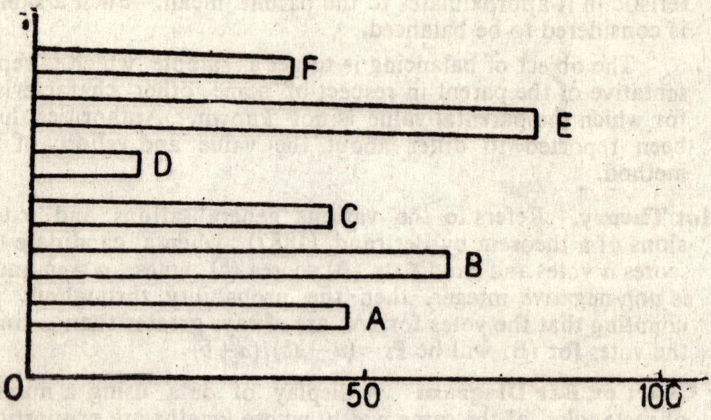

Fig. 2. Bar chart. Percentage increases in population, 1950-1980

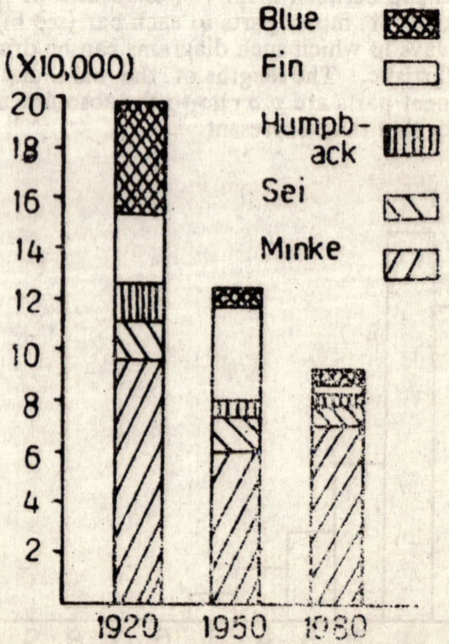

Fig. 3. Bar chart. Compound bar chart of estimated populations of five types of whale in an area of the Southern Ocean.

Bar Diagram. See Bar chart.

Bar Graph. A type of graphic presentation of data in which either horizontal or vertical bars are drawn to represent a distribution of cases. The bars are drawn to scale and represent proportionately the quantities being compared. The bar graph is most often used to portray the frequency distribution of a qualitative variable, with each bar representing a category of the variable, its height or length indicating the number of cases in that category. Bar graphs are also used to portray time series, with the height or length of each bar indicating the frequency or magnitude of the variable in each time period. They may also be used with quantitative variables, with the height or length indicating the number of cases in each class interval.

Bar Graph, Divided. A bar graph in which the bars are divided into segments that show the number of cases in two or more subdivisions of each category. For example, if the bars in a graph represented the categories of the variable marital status, each might be divided into two segments representing males and females, thus showing the proportion of each sex in sach category as well as the total distribution of cases among the categories.

Bartlett-Lewis Model. It is a stochastic cluster model in which the cluster-member process is considered to be a segment of a renewal process, which gets initiated by the cluster 'centre', and gets terminated after some finite number of 'renewals'.

Bartlett's Test. It is an approximate test for the homogeneity of a set of variances from a number of independent Normal samples. This test was given by Bartlett in 1937.

Base A number or magnitude which is used as a standard of reference, It may be found as a denominator in a ratio or percentage calculation. It may also be the magnitude of a particular time series from which a start has to be made in the calculation of a new relative series—an index number—which will show the observations as they get accrued in the future in relation to that of the base period.

Base Line. It is the horizontal line on a graph which is corresponding to some convenient basic measurement of the variable represented on the ordinate scale. The base is generally taken to be zero.

Base May. A map that shows the location of important geographic and economic features of an area, and is used to plot the distribution of social phenomena, such as income levels. It is also called social base map.

Base Period. Refers to the period of time for which data used as the base of an index number, or other ratio, are collected. This period is usually one of a year but it may be as short as one day or as long as the average of a group of years. The length of the base period will be governed by the nature of the material under review, the purpose for which the index number (or ratio) is being compiled and the desire to use a period as free as possible from abnormal influences so as to avoid bias.

Base Weight. Refers to the weights of a weighting system for an index number which are computed according to the information relating to the base period instead, for example, of the current period.

It is general in writing formulae to denote the information from the base period with a suffix *o* and that for the given period with a surfix *n*. A price index which is weighted according to prices and quantities of the base, *i.e.*, base weights, might be

$$I_{on}=\frac{\Sigma p_o q_o \left(\frac{P_n}{P_o}\right)}{\Sigma p_o q_o}$$

where p_o, q_o denote the base weights and summation is taken over the commodities composing the index.

Basic Statistics. The raw materials of statisticians ; usually collated, but rarely processed beyond the stage of correction for extraneous factors, such as differences in materials prices and wage rates where the basic statistics consist of annual costs. They can be classified in various ways, one being the dichotomy of continuous series and discrete series ; another dichotomy is that of time series and cross-section series, and a third, that of internal and environmental statistics.

Battery of Tests. A term which is used in applied psychology for a group of tests to which subjects get submitted. The usual objects of subsequent analysis have been either to provide predictions of each subject's aptitude for one or more occupations on the basis of a weighted combination of series in the tests, or else to provide variables whose correlations may subsequently get analysed into group or general factors common to the tests.

Bayes' Estimation. Refers to the estimation of population parameters by the use of methods of inverse probability and in particular of Bayes' theorem. If $\Pr(\theta/H)$ represents the prior probability of θ then the posterior probability of θ will be given by :

$$\Pr(\theta/x_1,x_2, \ldots, x_n,H) =$$
$$\Pr(\theta/H)\ \Pr(x_1,x_2. \ldots, x_n/\theta H)$$

θ is estimated by selecting that value which maximises the posterior probability. If Bayes' postulate gets invoked $\Pr(\theta/H)$ is constant and the method will be equivalent to the maximisation of the likelihood $\Pr(x_1, x_2, \ldots, x_n/\theta H)$.

Bayesian Controversy. Refers to a controversy between two schools of statistical thought. In Bayesian statistics, Probability is regarded as a numerical degree of belief. In some situations, like throwing a pair of dice, the belief can be supported objectively. In others, such as predicting the price of gold next week, any probability used has been essentially subjective. In non-Bayesian statistics, probabilities must rely on a frequency-based interpretation.

Statistical inference has been involving drawing conclusions about the situation underlying experimental data, and this may involve the use of Bayes' theorem on conditional probability.

The controversy arises because a non-Bayesian statistician fails to accept the validity of using this theorem if the probabilities are not frequency-based.

Bayes' Postulate. Refers to a postulate which is concerning the prior probabilities of a set of hypotheses (cf. Bayes' theorem) to the general effect that in the absence of information to the contrary all prior probabilities are to be considered equal. The postulate (or something equivalent) has been necessary if Bayes' theorem has to be applied in situations where there has been no information concerning the prior probabilities.

Bayes' Risk. Refers to the minimised value of the expected risk in a Bayes' solution.

Bayes Solution. In the terminology of statistical decision-functions, a Bayes' solution refers to a decision function which is able to minimise the average risk relative to some probability distribution.

Bayes' Strategy. Refers to one for which the available decision rules could be linearly ordered in terms of the Bayes' risk.

Bayes' Theorem, *n.* (On conditional probability), the theorem which states that

$$P(A \mid B)=\frac{p(B \mid A)\ p(A)}{p(B \mid A)\ p(A)+p(B \mid A')\ p(A')}$$

where $p(A \mid B)$ refers to the probability of event A occurring, given that event B has occurred, and where A′ is the event 'not' A′. Hence Bayes' theorem relates $p(A \mid B)$ to $p(B \mid A)$ and $p(B \mid A')$.

Bayes' theorem can be written in more general form as

$$p(A_r \mid B)=\frac{p(B \mid A_r)\times p(A_r)}{\sum_{r=1}^{r=n} p(B-A_r)\times p(A_r)}$$

where $A_1 \ldots A_n$ are mutually Exclusive and Exhaustive events. The theorem enables the probabilities of the explaining hypotheses to be determined ; this use is called the method of inverse probability.

The principal difficulty lies in determining, or even defining, the prior probabilities and the resolution of the difficulty by Bayes' postulate has occasioned much controversy. This does not, however, affect the theorem, which is a simple consequence of the product law of probability.

Bayesian Inference. Refers to a type of inference which regards parameters as being random variables possessed of prior distributions reflecting the accumulated state of knowledge. In its methods it is making use of Bayes' theorem and its extensions.

Bayesian Intervals. It is a method of summarising the information which is contained in the posterior probability distribution by quoting one or more intervals which contain certain amounts of probability. When such intervals have been based on a non-informative prior probability distribution they are sometimes termed as credible intervals (Edwards et al., 1963) or Bayesian confidence intervals (Lindley, 1965). When the posterior distribution has been symmetric, *i.e.*, that the tails have equal densities, then these intervals are termed as highest posterior density intervals because every point included is having a higher probability than those excluded.

Bayesian Probability Point. It is a critical value which is analogous to a confidence point or limit and derived according to the philosophy of Bayes' theorem and the use of the prior distribution (*s*) of the parameter (*s*) under estimation.

Behavioural Dicision Function. In statistical decision theory, it is a rule which is used for choosing between the possible decision functions, based on the outcome of an experiment.

Bell-Shaped Curve. It is a symmetrical frequency curve, which is usually of a continuous frequency distribution, and shows a marked similarly to a vertical section through a bell.

Bellman-Harris Process. It is a branching process which evolves from an initial individual, in which each individual lives for a random length of time and at the end of its life produces a random number of offsprings of the same type.

Belt Graph. A type of segmented line graph in which several line graphs, combined in a single chart, show the component elements being depicted as well as their total. The belt graph is the linear counterpart of the divided bar graph. It shows the distribution of the component parts individually and together.

Bernoulli Distribution. It is another name for Binomial Distribution. The term is, however, generally restricted to the special case of the binomial distribution when n=1.

Number of successes :	0	1
Probability :	$1-p$	p

Thus it is the probability distribution of the number of successes in a single trial.

Bernoulli Numbers. The Bernoulli number of order r, B_r, is defined as the numerical coefficient of $t^r/r!$ in the expansion of $t/(e^t-1)$ as a power series in x.

Explicitly, $B_0=1$, $B_1=-\frac{1}{2}$, $B_2=\frac{1}{6}$, $B_4=-\frac{1}{30}$, $B_6=\frac{1}{42}$, $B_3=B_5=B_7=\ldots=0$.

There are variations in the notation and some writers give

$B_1=-\frac{1}{2}$, $B_2=\frac{1}{6}$, $B_3=-\frac{1}{30}\ldots$.

Bernoulli Polynomial. The Bernoulli polynomial $Br^{(n)}(x)$, of order n and degree r is defined as the coefficient of $t^r/t!$ in the expansion of

$$\left(\frac{1}{e^t-1}\right)^n e^{xt}.$$

Bernoulli Process. A process in which each trial has only two possible outcomes, the probability of the outcome of any trial remains fixed over time and trials are statistically independent.

Bernoulli's Theorem. A theorem propounded by James Bernoulli in fourth part of his *Ars Conjectandi* which was published in

1713 after his death in 1705. It is the theorem which states that, if the probability of success in a trial is p, the probability of exactly r successes in n independent trials is.

$$nC_r p^r (1-p)^{n-r}$$

where $nC_r = \dfrac{n!}{(n-r)!r!}$ and $n!$ (n factorial) is

$1\times2\times3\times\ldots\times n$. (See also binomial coefficients.)

This is the theorem underlying the Binomial Distribution.

Effectively it is a proposition in pure mathematics to the effect that the observed proportional frequency in random drawings of individuals from a population of attributes with constant probability p converges to p in probability; or, to put it another way, the proportional frequency in the binomial distribution $(q+p)^n$ lying within a range $\pm\epsilon\sigma$ of the mean value p tends to unity with increasing n however small ϵ may be, σ being the standard deviation $\sqrt{(pq/n)}$.

Bernoulli himself seems to have regarded the theorem as something beyond a mathematical proposition, perhaps a justification of a frequency theory of probability.

Bernoulli Trial. It is an experiment with fixed probability p of success, $1-p$ of failure. In a sequence of Bernoulli trials, the probability of success remains constant from one trial to the next, independent of the outcome of previous trials.

The probability distribution for the various possible numbers of suceesses, 0 to n, in n Bernoulli trials has been the binomial Distribution. This can also be called the bernoulli distribution, but this term is often restricted to the case when $n=1$.

Bernoulli Variation. Refers to a sampling situation in which members are selected from a population of attributes such that the probability of occurrence remains constant, hence the sampling distribution of occurrence in samples of fixed size has been binomial. The term finds use in contradistinction to Lexis and Poisson variation.

Bernoulli Walk. Refers to a random walk on the integers, in which at each step the conditional probabilities of moving up or down by one unit have been

$$p_{i,i+1} = p, i = \ldots -1, 0, 1, \ldots;\ 0<p<1.$$
$$p_{i,i-1} = 1-p$$

Bessel Function Distribution. It is a frequency distribution which is involving Bessel functions. For example, the distribution of the covariance of two Normal correlated variates involves a Bessel function of the second kind having imaginary argument.

Bessel's Correlation. A factor by which the variance of a small sample is multiplied to provide an estimate of the variance of the universe. The correction factor is

$$\frac{n}{n-1}$$

where n is the number of items in the sample. The greater the value of n. the closer the correction approaches to unity, so that for very large samples the correction becomes unnecessary. See sampling.

Best Estimator. The estimation of population parameters from information provided by the sample gives rise to the question whether there has been a 'best' estimasor. The answer has been found to depend mainly on the criteria which are laid down as to the 'goodness' of an estimator. If there has been a criterion which distinguishes one of two estimators as better than the other and if there exists an estimator which is better than other, it is considered to be the best.

Various criteria are suggested, *e.g.* that of sufficiency, minimum variance or closeness. It is not always true that a 'best' estimator exists.

Best Fit. See goodness of fit.

Best Fit, Line of. A straight line or curve fitted as a graph representing the relationship between two variables plotted one against the other in a scatter diagram. Figure shows a scatter diagram of monthly costs plotted against monthly output, The line of best fit may be drawn freehand by personal judgement, as it is in Figure ; or it may be fitted by group averages or least squares, or by any other method of simple-regression analysis. When the line has been drawn, then if it turns out to be straight, readings can be taken from it in order to derive its equation. If it turns out to be a curve, then it would be necessary for analytical purposes to determine the type of equation to which it conforms, *e.g.* logarithmic or quadratic. See model building.

Best Linear Unbiassed Estimator. It is a linear function of the order statistics which is available in the sample of information of the variable under analysis for which the variance has been minimum and which has been also Unbiassed.

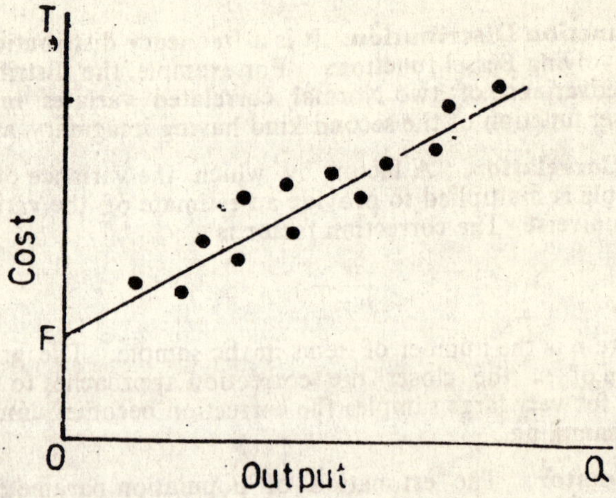

Scatter diagram showing lives of best fit.

Beta. The probability of a type II error.

Beta-binomial Distribution. Refers to a discrete probability distribution of the form given below :

$$p(r)=\binom{n}{r}\frac{B(\alpha+r,\ n+\beta-r)}{B(\alpha,\ \beta)} \quad r=0,\ 1,\ 2,\ \ldots..\ n.$$

It is the convoluted distribution which is obtained by averaging the probability parameter of a binomial distribution having index n over a beta distribution having parameters α and β.

Beta Coefficient (β). A standardized measure of a regression coefficient (R8) ; achieved by 'stating each of the variables in units of their own individual standard deviation'. A simple regression equation (R9) may be written ;

$$Y=aX+b$$

Its beta equivalent is :

$$\frac{Y}{\sigma y}=\beta\frac{X}{\sigma x}+b^1$$

Each figure of the Y series is divided by the standard deviation of the Y series, and each figure of the X series is divided by the standard deviation of the X series, before the regression analysis is carried out. The standard deviation of a series expressed in units of cwt is 20 times that of the same series expressed in units of a ton ; the resultant ratios are the same whether the series is expressed in cwts or tons. At the same time the size of any regression coefficient depends partly on the units in which the basic series are expressed. The beta coefficient eliminates this factor from the regression, and the several beta coefficients of a multiple-regression equation can be compared directly one with another, and with the beta coefficients of other regression equations.

Beta Distribution. Refers to the distribution with probability density function given by :

$$f(x)=\frac{x^{\alpha-1}(1-x)^{\beta-1}}{B(\alpha\beta)}$$

where the parameters $\alpha\beta>1$, $0\leqslant x\leqslant 1$, and $B(\alpha, \beta)$ is the *beta function* given by

$$B(\alpha, \beta)=\int_0^1 x^{\alpha-1}(1-x)^{\beta-1}\,dx$$

The beta distribution is having

$$Mean=\frac{\alpha}{(\alpha+\beta)}$$

$$Variance=\frac{\alpha\beta}{(\alpha+\beta+1)(\alpha+\beta)}$$

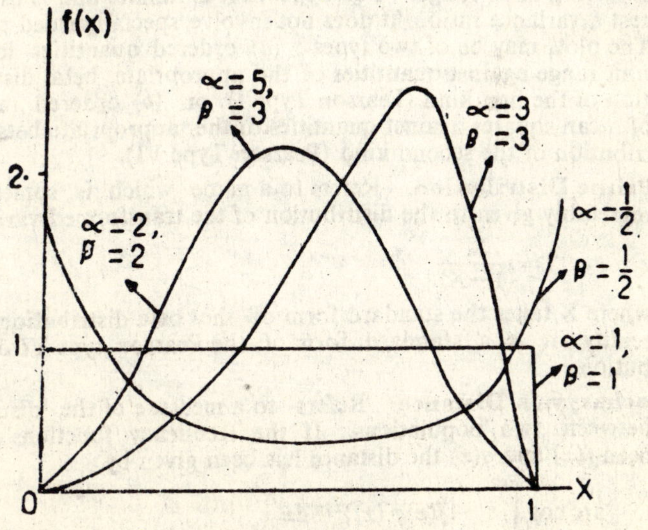

Fig. 1. Beta Distribution.

The graph of the beta distribution get varied in shape according to the values of α and β as shown in Fig. 1.

A second form, sometimes called a beta distribution of the second kind, is

$$dF=\frac{y^{\alpha-1}}{B(\alpha,\ \beta)(1+y)^{\alpha+\beta}}\,dy,\ 0\leqslant y\leqslant\infty\ ;\ \alpha,\ \beta>0.$$

This easily gets transformed into the first type by putting $x=1/(1+y)$. It is also referred to as an 'inverted' beta distribution.

Distributions of the first kind have been a special case of the Pearson type I distribution and those of the second kind are a special case of the Pearson type VI distribution.

β-Error. In the theory of testing hypotheses, it refers to an error which is incurred by accepting an hypothesis when it is incorrect. This kind of error is also termed as consumer's risk, an error of second kind or a Type II error.

Beta Function. See beta distribution.

Beta Probability Plot. Refers to the graph of cumulated distribution function as ordinate against the variate as abscissa for beta (Type I or Type VI) distributions. For Type I the variate is having finite range. For Type VI it is infinite and is used to test a variance ratio. It does not involve specially ruled paper. The plots may be of two types : (*a*) ordered quantities in the unit range against quantities of the appropriate beta distribution of the first kind (Pearson Type I) or (*b*) ordered rations of mean squares against quantities of the appropriate beta distribution of the second kind (Pearson Type VI).

Beta-Prime Distribution. Refers to a name which is sometimes regretably given to the distribution of the transformed variable

$$T=\frac{X}{(1-X)}$$

where X takes the standard form of the beta distribution. In reality, it is a standard form of the Pearson type VI distribution.

Bhattacharyya's Distance. Refers to a measure of the 'distance' between two populations. If the frequency functions have been $f(x)$ and $g(x)$ the distance has been given by

$$\text{arc cos}\int_{-\infty}^{\infty}\{f(x)g\ (x)\}^{1/2}\,dx$$

Analogous expressions could be given for discontinuous and multivariate distributions.

Bias. Refers to a systematic error in a statistical result. Errors from chance will cancel each other out in the long run, but those from bias will not.

Generally, bias has been an effect which deprives a statistical result of representativeness by *systematically* distoring it, as distinct from a random error which may distort on any one occasion but balances out on the average.

Biased Estimator. A formula or procedure which produces a biased estimate of a Parameter of the parent population from sample data. In statistics it is common to estimate properties of a pasent population, like its Mean and Variance, by examining a sample. A formula or procedure which predicts a property or parameter of the parent population using the data is called an estimator. If its expected value over all possible samples is equal to the quantity it estimates, it is unbiased. Otherwise it is biased.

If $x_1, x_2, \ldots, x_n$ are the values of a random variable measured, for a sample of size x, then

$$\frac{x_1+x_2+\ldots+x_n}{n}=\bar{x}$$

The sample mean x is an ***unbiased estimator*** for the parent mean.

The sample variance is, however, a biased estimator for the parent variance :

$$\sum_{i=1}^{n}\frac{(x_i-\bar{x})^2}{n}$$

An unbiased estimator for it is :

$$\sum_{i=1}^{n}\frac{(x_i-\bar{x})^2}{n-1}$$

Also, see unbirsed estimator.

Biased Sample. Refers to a sample which is obtained by a biased sampling prosess, that is to say, a process which incorporates a systematic component of error, as distinct from random error which balances out on the average. Non-random sampling is gencrally, though not inevitably, subject to bias, particularly when entrusted to subjective judgement on the part of human beings.

Biased Test. A test is considered to be biased if it provides a lewer probability of rejecting the hypothesis under test (H_0) when the alternative hypothesis (H_1) has been true then when H_0 has been true. Expressed in another way if the hypothesis under test has been $\theta=\theta_0$ and the power function of the test has a minimum value at a point $\theta\neq\theta_0$ then the test has been considered to be biased.

Bimodal (of a Distribution) 1. Having two Modes. For example, the goals scored per match by a football team in a season are given in the table below. As there have been two modes, 0 and 2, so the distribution has been bimodal.

Goals/match	*Frequency (matches)*
0	12
1	8
2	12
3	6
4	3
5	1

2. A distribution having two distinct, but unequal, local modes, a shown below :

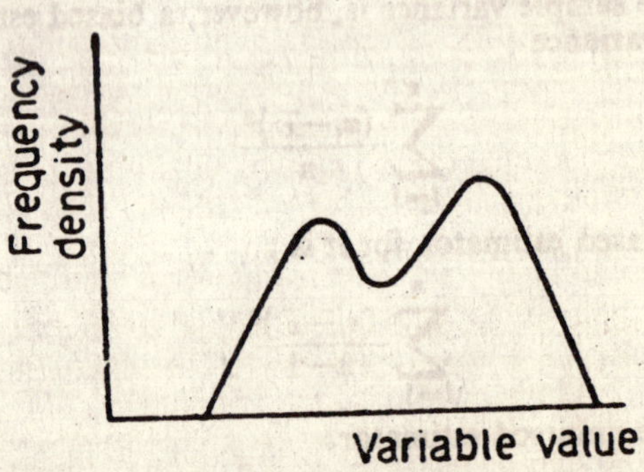

A bimcdal distribution.

Bimodal Distribution. Refers to a distribution of data points in which two values occur mcre frequenly than the rest of the values in the data set.

Binary Numbers. Decimal numbers consist of the ten numbers 0 to 9; binary numbers, of the two numbers 0 and 1. In binary, then, 10 equals 2 of the decimal notation, and is described in terms of the decimal system as being of the *base* 2; the decimal system is described in its own terms as being of base 10. The following table of equivalents is a useful and revealing guide to conversion.

Rinary	*Decimal*
1	1
10	2
100	4
1,000	8
100,00	16
100,000	32

By addition:

11	3
101	5
110	6
111	7
1,001	9
1,010	10
1,011	11

It will be seen that, as the binary numbers shown proceed from 1 upwards in powers of 10 (binary), the corresponding decimal numbers proceed in powers of 2 (dcemial), *i.e.*, 2^0, 2^1, 2^2, 2^3,....

Computers use the binary system internally for technical reasons. Print-outs are invariably stated in decimal numbers.

Binary Sequence. In general, it refers to any sequence each numbers of which can take one of two possible values. In a probabilistic context it refer to a series of numbers each of which represents the outcome of an (independent or dependent) binomial trial and hence where the two possible out-comes might be represented as 0 or 1.

Binomial Coefficients. Refers to the coefficients of the terms in the expansion of $(x+y)^n$ for positive integer values of n. Symbol: ${}^nC_r$.

Example :

$$(x+y)^5=1x^5+5x^4y+10x^3y^2+10x^2y^3+5xy^4+1y^5$$

and so the binomial coefficients for $n=5$ would be 1, 5, 10, 10, 5, 1.

Two common notations have been used for writting these :

(*a*) $^5C_0$, $^5C_1$, $^5C_2$, $^5C_3$, $^5C_4$ and $^5C_5$

(*b*) $\binom{5}{0}$, $\binom{5}{1}$, $\binom{5}{2}$, $\binom{5}{3}$, $\binom{5}{4}$ and $\binom{5}{5}$.

Two ways of working out binomial coefficients—are known :

(*a*) By the formula $^nC_r=\dfrac{n!}{r!(n-r)!}$

Hence $^5C_2=\dfrac{5!}{2!3!}=\dfrac{5.4.3.2.1}{2.1.3.2.1}=10$

(*b*) By using Pascal Triangle :

```
                1
              1   1
            1   2   1
          1   3   3   1
        1   4   6   4   1
      1   5  10  10   5   1
    1   6  15  20  15   6   1
  △   △   △   △   △   △   △   △
```

Relationships involving binomial coefficients include

$$^nC_r={}^nC_{n-r}$$

$$^nC_r+{}^nC_{r+1}={}^{n+1}C_{r+1}$$

$$^nC_0+{}^nC_1+{}^nC_2+\ldots+{}^nC_n=2_n$$

^nC_r also gives the number of ways of selecting r objects from n, without regard to their order.

Binomial Distribution. The theoretically expected distribution of a series of random samples drawn from a population with two categories or classes. A binomial distribution would show the theoretical frequency with which one would expect to obtain random samples with each of a series of proportions of cases in one of the two categories. For example, if 50 random samples were drawn from a population half male and half female, the binomial distribution could give the theoretically expected number of samples that would be all made, 90% male, 80%

male, 70% male, etc. The binomial distribution is one of the important mathematical models used in statistics. See poisson distribution ; sample, random.

Binomial distribution is also known as the Bernoulli distribution after James Bernoulli who give it in his (posthumous) *Ars Conjectandi* in 1713.

If the probability of success is p, and that of failure $q=1-p$, then the probability of r successes and $n-r$ failures is given by :

$$^nC_r p^r q^{n-r} \left(\text{the binomial coefficient } ^nC_r = \frac{n!}{r!(n-r)!} \right)$$

The probability generating function for the binomial distribution has been given by $(q+pt)^n$ since, when expanded, this provides ;

$$^nC_0 q^n + {}^nC_1 q^{n-1} (pt) + \ldots + {}^nCrq^{n-r}(pt)^r + \ldots {}^nC_n (pt)^n$$

It is often convenient to write the expression to be expanded as $(pX+pY)^n$, where event X is having probability p, event Y, *i.e.*, X′, or not X, has probability $q=1-p$. X and Y can then be thought of as markers, so that the probability of X occurring r times, and Y occurring $n-r$ times, is given by the coefficient of X^rY^{n-r}, namely :

$$^nC_{n-r} p^r q^{n-r} = \frac{n!}{(n-r)!r!} p^r q^{n-r}$$

The mean, or expectation, of the binomial distribution is given by np, the standard deviation by $\sqrt{npq}$.

The distribution of the proportion of successes in n trials is binomial with

$$\textit{Mean} = p$$

$$\textit{Standard deviation} = \sqrt{\frac{pq}{n}}$$

Binomial Index of Disperison. A statistic which is used for testing whether a set of samples has been homogenous with respect to some common attribute.

If there have been k samples of sizes $n_1, \ldots n_k$ with proportions $p_1, \ldots, p_k$ and p has been the mean proportion for all members together; *i.e.*,

$$p=\left(\sum_{i=1}^{k} n_i p_i\right) \Big/ \sum_{i=1}^{k} n_i,$$

the index of dispersion is

$$\sum_{i=1}^{k} n_i\,(p_i-p)^2\{p(1-p)\}.$$

The significance of the index, as denoting departure from homogeniety, may be tested in the χ^2 distribution having $k-1$ degrees of freedom.

The index has been a particular case of the Lexis ratio.

Binomial Probability Paper. A graph paper having a gird which is specially designed to facilitate the analysis of enumeration data, *i.e.*, data in the form of proportions or percentag from a binomial population. Both the rectangular coordinate axes have been graduated in terms of the square-root of the variable.

Binomial Test. A nonparametric statistical test that may be used when research data are divided into just two categories (*e.g.*, male and female, favorable responses and unfavorable responses, over 21 years of age and under 21 years of age). The binomial test uses the Binomial Distribution to test whether the distribution of cases in the sample studied differs significantly from the distribution that would be expected if the sample were drawn randomly from a particular two-class population. It is a Goodness-of-Fit Test.

Binomial Variation. It is another name for Bernoulli variation.

Binomial Waiting Time Distribution. It is alternative name for the negative binomial distribution which refers to the realisation of a success for the kth time after $k+x-1$ independent binomial trials.

Biometry. Refers to the study of biological data by means of statistical analysis.

Birth and Death Process. It is a stochastic process which tries to describe the growth and decay of a population the members of which may die or give birth to new individuals. The types mainly studied passess relatively simple laws of reproduction and mortality.

Birth, Death and Immigration Process. It is a simple extension of birth and death process which takes into consideration immigration into the system out-side. Immigration could be treated synonymously with 'death'.

Birth Process. It is a stochastic process which describes the population of a system in which individual members may give birth to new members. The expression is generally confined to the case where the variate (population) increases only by jumps of amount+1, the probability of a jump from n to $n+1$ in time dt being asymptotically $\lambda_n dt$. Here λ_n may also be depending on t.

Birth Rate. The crude birth rate of an area may be defined as the number of births actually occurring in that area in a given time period divided by the population of the area as estimated at the middle of the particular time period. The rate has been usually expressed in terms of 'per 1000 of population'.

This crude birth rate has been capable of considerable refinement according to the various specific viewpoints adopted in the analysis of vital statistics. For example, it may get adjusted to allow for changes in the proportion of the female population in the child-bearing age groups.

Biserial. (Of the relationship between two variables) having one variables that takes only two values.

Bit. An abbreviation for binary digit. It is a digit of a number which is written in the scale of two ; for example, the number 2 is expressed as 10, 4 as 100 and 8 as 1000.

Bivariate Beta Distribution. It is a bivariate continuous distribution which is due to Filon and Isserlis (K. Pearson, 1923), having the joint probability density

$$h(x, y)=\frac{\lambda\,(p_1+p_2+p_3)}{\lambda\,(p_1)\,\lambda\,(p_2)\,\lambda\,(p_3}\,x^{p_1-1}y^{p_2-1}(1-x-y)^{p_3-1}$$

where $x\geqslant 0$, $y\geqslant 0$, $x+y\leqslant 1$ and $p_1, p_2, p_3>0$. The distribution can be derived as the joint distribution of

$$X=X_1/(X_1+X_2+X_3),$$
$$Y=X_2/(X_1+X_2+X_3)$$

where X_1, X_2, X_3 denotes independent random variables having gamma distributions with index parameters p_1, p_2, and p_3.

Bivariate Binomial Distribution. It is an extension of the binomial distribution to the case where a member can exhibit success or failure in each of two attributes. If the probabilities of success for the attributes have been denoted by p_{11}, p_{10}, p_{01} and p_{00} the probability of x successes of the first and y of the second in a sample of x is given by

$$\binom{s}{x}\sum_{i}\left\{\binom{s-x}{y-i}\binom{x}{i} p_{11}^{\,i}\; p_{10}^{\,x-i}\; p_{01}^{\,y-i}\; p_{00}^{\,s-x-y+i}\right.$$

where the summation occurs up to i=the smaller of x, y.

Bivariate Cauchy Distribution. A bivariate probability distribution having density function

$$f(x_1, x_2)=\frac{a}{2\pi\ (a^2+x^2_1+x^2_2)^{3/2}},\ -\infty<x_1, x_2<\infty,\ a>0.$$

Bivariate Data. Data in which there are two measurements for each item, for example, the income and years of education of each subject studied.

Bivariate Distribution. Refers to the distribution of a pair of variates x_1 and x_2, written for the continuous case as

$$dF=f(x_1, x_2)\ dx_1,\ dx_2.$$

For discontinuous or grouped data the distribution may be set cut in a rectangular array termed as a bivariate or correlation table. It is often used by statisticians as a preliminary centre to show whether two series are significantly correlated.

If the probabilities of the two. variables have been given rather than their frequencies, the distribution is then a *bivariate probability distribution.* The distributions obtained by adding along each row or column of a bivariate distribution table have been termed as *marginal distributions.*

In the example of the second-hand cars (Fig. 9), the marginal distributions are :

Mileage (×1,000)	0-10	10-20	20-30	30-40	40-50	50-60	60-70	70-80
Frequency	1	2	2	3	7	6	6	3

and

Age	0-1	1-2	2-3	3-4	4-5	5-6	6-7	7-8
Frequency	3	2	7	9	3	3	1	2

If each row (or column) of a bivariate distribution table has the same variance, the distribution is said to be *homoscedastic* with respect to that variable, otherwise *heteroscedastic.*

Bivariate Exponential Distribution. It is any bivariate distribution that is having exponential marginal distributions. Some have been special cases of bivariate gamma distribution which are obtained by setting the index parameters to unity. Others include Gumbel's (1960) bivariate exponential distribution with probability density function

$$h(x, y)=\{(1+ax)\ (1+ay)-a\}\ e^{-x-y-axy}$$

$$x\geqq 0,\ y\geqq 0,\ a\geqq 0,$$

and the bivariate case of the Marshall-Oklin distribution.

Bivariate-F-Distribution. It is a bivariate generalisation of the F-distribution which arises naturally in the simultaneous testing of linear hypotheses in the analysis of variance. The distribution is having probability density function

$$h\ (x, y)=\lambda \left[\frac{v}{2}\right]\ v_0^{-(1/2)v} \prod_{i=1}^{2} \{v_1^{\ v i/2/1}\ (\tfrac{1}{2}v_i)\}$$

$$=\frac{x^{(v_1-1)/2}\ y^{(v_2-1)/2}}{\left(1+\dfrac{v_1x+v_2y}{v_0}\right)^{v/2}}\quad 0\geqq x,\ y<\infty$$

where $v=v_0+v_1+v_2$, which can be shown to be the joint density of

$$X=\frac{X_1/v_1}{X_0/v_0}\quad \text{and}\quad Y=\frac{X_2/v_2}{X_0/v_0}$$

where X_0, X_1, and X_2 represent independent central chisquared varaites with v_0, v_1 and v_2 degrees of freedom respectively (Ghosh, 1955).

Bivariate Gamma Distribution. Refers to any bivariate distribution having gamma marginal distribution. Examples of such distributions are cited by Wicksell (1933), McKay (1934) and Cherian (1941).

Bivariate Hypergeometic Distribution. It is a discrete bivariate distribution having joint probability mass function

$$\Pr\ (x, y)=\frac{\binom{Np_1}{x}\begin{bmatrix}Np_2\\ y\end{bmatrix}\begin{bmatrix}N-Np_1-Np_2\\ N-x-y\end{bmatrix}}{\begin{bmatrix}N\\ n\end{bmatrix}}$$

where $x, y=0, 1,\ldots, n$ such that $x\leqq Np_1$, $y\leqq Np_2$ and $n-x-y\leqq N\ (1+p_1-p_2)$; N, n have been integers with $n\leqq N$; $0<p_1$, $p_2<1$ and $p_1-p_2\leqq 1$. The distribution, which was introduced by Isserlis (1914), is having hypergeometric conditional distributions $\Pr\ (x \mid y)$ and $\Pr\ (y \mid x)$.

Bivaraite Logarithmic Distribution. It is discrete bivariate distribution which was introduced by Khatri (1959). It is having joint probability mass function.

$$\Pr(x, y)=\frac{(x+y-1)!}{x!\,y!}\cdot\frac{p_1^x p_2^y}{-\log(1-p_1-p_2)}$$

where $x, y=0, 1,\ldots;\ x+y\geqq 1,\ 0<p_1, p_2<1$, and $p_1-p_2<1$, and arises as the limit of the (0, 0)-truncated bivariate negative binomial distribution. Alternatively it is termed bivariate logarithmic series distribution.

Bivariate Negative Binomial Distribution. It is descrete bivariate distribution having joint probability mass function.

$$\Pr(x, y)=\frac{(x+y+k-1)!}{x!y!k}\,p_1^x p_2^y(1-p_1-p_2)^k$$

where $x, y=0,1,\ldots;\ k>0,\ 0<p_1,\ p_2<1$, and $p_1+p_2<1$. The conditional distributions $\Pr(x \mid y)$ and $\Pr(y \mid x)$ have been negative binomial.

Guldberg (1934) was first of all introduced this distribution, and Lundberg (1940) and first used it in connection with problems of accident proneness.

Bivariate Normal Distribution. Two variates x_1 and x_2 having means m_1 and m_2 and variances σ_1^2 and σ_2^2 get distributed in the bivariate Normal form if their distribution function has been given by

$$dF=\frac{1}{2\pi\sigma_1\sigma_2(1-p^2)^{1/2}}\exp\left[-\frac{2}{2(1-p^2)}\left\{\left(\frac{x_1-m_1}{\sigma_1}\right)^2-\frac{2p(x_1-m_1)(x_2-m_2)}{\sigma_1\sigma_2}+\left(\frac{x_2-m_2}{\sigma_2}\right)^2\right\}\right]dx_1dx_2,$$

where p denotes a parameter not greater than unity in absolute value. For any fixed x_1 (or x_2) the variate x_2 (or x_1) is Normally distributed. The parameter p has been (or ought to be) called the correlation parameter.

Bivariate Pascal Distribution. Suppose an experiment of observing individuals is considered one after another for two characters A_1 and A_2 until exactly k individuals having both the characters are observed. If x_i denote the number of individuals possessing character A_i observed in the experiment, $i=1, 2$, the x_1, x_2) has the bivariate Pascal distribution. This distribution has been a special case of the bivariate negative binomial distribution.

Bivariate Poisson Distribution. It is discrete bivariate distribution having frequency function of the form $f(x, y\,;\,\theta)$

$$=\exp\,(\theta-\lambda_1-\lambda_2)\sum_{k=0}^{\min(x,\,y)}\frac{\theta^k(\lambda_1-\theta)^{x-k}(\lambda_2-\theta)^{y-k}}{k!(x-k)!(y-k)!}\,.$$

$$x=0,\,1,\,2\ldots\,;\;y=0,\,1,\,2,\ldots$$

where $0\leqslant\theta\leqslant\min\,(\lambda_1,\lambda_2)$. The marginal distributions of the random variables X and Y have been Poisson with parameters λ_1 and λ_2 respectively.

Bivariate 'Student' Distribution. It is a bivariate generalisation of 'Student's t-distribution' which was introduced by Pearson (1923) for graduation purposes. The distribution is having the probability density function

$$h(x,\;y)=\{2\pi\sigma_1\sigma_2\sqrt{(1-p^2)}\}^{-1}\;\frac{f}{f-2}$$
$$\times\left[1+\frac{1}{(f-2)(1-p^2)}\left\{\frac{x^2}{\sigma_1^2}-\frac{2pxy}{\sigma_1\sigma_2}+\frac{y}{\sigma_2^2}\right\}\right]^{-(f+2)/2}$$

where $-\infty<x,\,y<\infty,\,j>0$. The random variables X/σ_1 and Y/σ_2 are having marginal t-distributions with f degrees of freedom, and X and Y are having product moment correlation coefficient p.

Bivariate Type II Distribution. It is a continuous bivariate distribution having probability density function

$$h(x,\;y)=\{2\pi\sigma_1\sigma_2\sqrt{(1-p^2)}\}^{-1}\,\frac{n+1}{n+2}$$
$$\times\left[1-\frac{1}{2(n+2)(1-p^2)}\times\left\{\frac{x}{\sigma_1^2}-\frac{2pxy}{p_1p_2}+\frac{y}{\sigma_2^2}\right\}\right]^n$$

where $-c\sigma_1<x<c\sigma_1,-\sigma_2<y<c\sigma_2,\;c^2=2(n+2)$ and $n>0$, introduced by Pearson (1923). The marginal distributions of both X and Y have been type II distributions.

Bivariate Uniform Distribution. It is any bivariate probability distribution that is having uniform marginal distributions. Examples have been the contingency-type uniform distribution, otherwise known as Plackett's uniform distribution, and Morgenstern's uniform distribution. (See under contingency-type distributions and under Morgenstern distributions.)

Blind Test. A technique for evaluating consumer's reactions to products, especially food, in which a group of people are given unidentified samples and asked for their comments.

Block. Refers to a subset of the items under investigation in a statistical experiment. It could be :

The barley in a particular part of a field.

The householders in one street of a town.

The weekly earnings of one particular employee of a factory.

The variability within a block would be expected to be less than that within the whole population being investigated.

For example, a block may comprise a group of contiguous plots of land, all the animals in a litter, the results obtained by a single operator, or meteorological observations on a series of days at a given place. The general purpose of dividing all the material in an experiment into blocks (or of regarding it as so divided by the circumstances of the case) has been to isolate sources of heterogeneity ; the items in a block being so far as possible homogeneous and uncontrolled variation in the experimental material being measured by comparisons between blocks.

The variation in the experimental observation has been usually divided (by variance-analysis) into effects due to differences between blocks and effects, due to variation within blocks ; these being known as interblock and intrablock effects (variances, etc.) respectively. The expressions 'interblock' and 'intrablock information' occur both in the general sense and in the specialised sense of information. Thus for estimates of certain treatment comparisons the 'information' may be the reciprocal of the most efficient estimator if it exists.

Block Diagram. It is made up of vertically placed rectangles which are situated adjacent to each other on a common base line. Where the characteristic to be depicted has been quantitative the height of the rectangles is generally considered to be proportional to this quantitative variable. When this kind of diagram is used to portray a frequency distribution it takes the name of histogram.

Boole's Inequality. An inequality which was developed by Boole in 1854 to provide the limits to the frequencies in certain logically defined classes in terms of the frequencies in other classes. It finds application in probability theory. One simple form states that if $A_1, A_2, \ldots, A_k$ have been compatible events, the probability that at least one occurs will not be greater than the sum of the probabilities that each occurs independently of the occurrence of the others :

$$1-\Pr(\text{not-}A_1, \text{not-}A_2, \ldots, \text{not-}A_k) \leqslant \Pr(A_1)+\Pr(A_1)+\Pr(A_2)+\ldots+\Pr(A_k).$$

Borel-Tanner Distribution. In queueing theory it refers to the distribution, under certain conditions, of the number of customers served before the queue gets vanished for the first time.

Bose Distribution. Refers to the distribution of the variance ratio when the variates get correlated. It was given by R.C. Bose (1935).

Bose-Einstein Statistics. In statistical mechanics one possible basic assumption concerning states and energy levels in equivalent to supposing that r distinguishable particles get distributed among n cells $(r<n)$ in such a way that each of the n^r arrangements has been equally probable. This results in the Maxwell-Boltzmann statistics. If the particles have been indistinguishable there have been $\binom{n+r-1}{r}$ distinguishable arrangements and if these are taken as equally probable there result Bose-Einstein statistics. As a particular case, if not more than one particle may appear in any cell there are $\binom{n}{r}$ equally probable arrangements ; and these are forming the basis of Fermi-Diras statistics.

Box Three Boxes Game. A game in which there are three boxes marked 1, 2, and 3. For a given play of the game, player A removes the bottom of one of the boxes, but player B does not know which one it is. Player B then puts an amount of money equal to the number marked on the box in each of two of the three boxes. He loses the money put in the box with no bottom and wins the money put in the others. This is a *zero-sum* game with *imperfect information.* The *payoff matrix* does not have *a saddle point* and the solutions are *mixed strategies.* The solutions are $(0, \frac{1}{2}, \frac{1}{2})$ for A and $\left(\frac{3}{5}, \frac{2}{5}, 0\right)$ for B, meaning that A removes the bottoms of boxes 2 and 3, each with probability $\frac{1}{2}$, player B puts money into boxes 1 and 2, or 1 and 3, with respective probabilities $\frac{3}{5}$ and $\frac{2}{5}$ (never in 2 and 3).

The *value* of this game is 1 (with B the *maximizing player.*)

Box-and-Whisker Plot. A diagram which is used to display some important features of the distribution of experimental data (see exploratory data analysis). It consists of two lines (whiskers) drawn from points representing the extreme values of the lower and upper quartiles, and a box drawn between the quartiles which include a line for the median. A box-and-whisker plot is also termed as a *boxplot.*

The boxplot is drawn as in Fig. The box, drawn between the lower and upper quartiles, contains 50% of the date. The two Wiskers, from the quartiles to the least or greatest values, each contain 25% of the data.

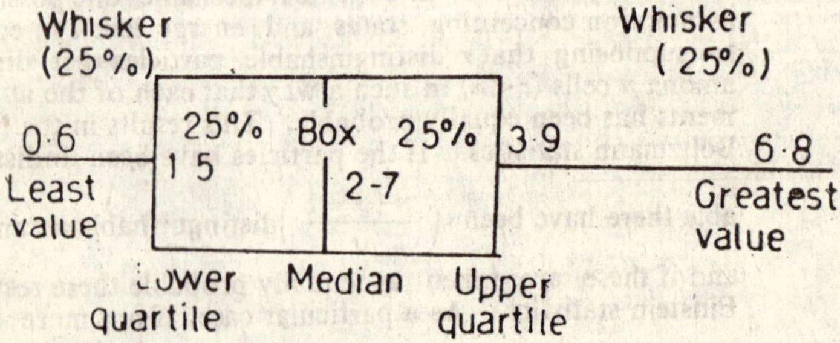

Box-and-whisker plot.

Boxplot. See box-and-whisker plot.

Box's Test. It is an approximate test which is used for the equality of the variances of a number of populations. The important feature of the test has been its robustness under departures from Normality in the populations.

Box-Jenkins Model. It is a model which is used for predicting non-stationary time series with or without seasonal variations. It has been found to depend on autoregression of the *m*th difference of the series, having a moving average residual.

Branching Process. It is a stochastic process which describes the growth of a population in which the individual members may have offspring, the lines of descent 'branching-out' as new members have been born. Sometimes it is referred to as a chain-reaction process. If the life-time distribution of the individual members, *i.e.*, the waiting-time in a particular state, follows the negative exponential form the (continuous time) process is called a branching Markov process. For a general form of life-time distribution the term age dependent branching process has been employed used.

Branching Renewal Process. It is a renewal process in which each of a series of primary events generates a subsidiary series of events. The complete process involves the superposition of the events in the primary and subsidiary processes.

Bravais Correlation Coefficient. It is an obsolete synonym for the correlation parameter in bivariate Normal variation ; and, by extension, to the productmoment correlation coefficient which estimates it.

Breakdown Bound. Refers to one of a member of similar measures (Hampel, 1971) which are used to compare the robustness of estimators in the presence of extreme outliers. The bound provides an indication of the largest proportion that an estimator may be expected to tolerate.

Break Even Analysis. A technique to determine the rate of sales in quantity so that the rate of money sales can be equated to the rate of costs ; usually expressed in terms of annual rates. There are two approaches : one, the linear, in which it is assumed that the selling price of the product remains constant, differences in the rate of sales being due to other factors ; the other, the curvilinear, in which it is assumed that differences in the rate of sales are due to differences in the selling price.

Linear approach. Figure 1 demonstrates this. The cost graph, T, is a straight line intersecting the vertical axis at a point, F, which is the measure of the fixed annual costs (F8), and S is the sales graph, which is also a straight line since the price is held constant for all rates of sales. The point of inter-

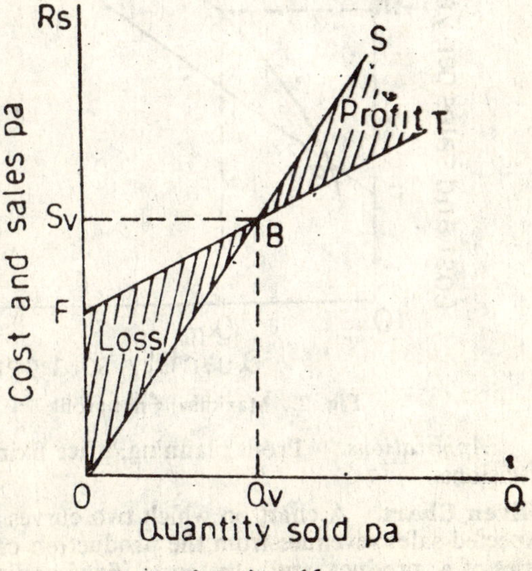

Fig. 1. Quantity sold per an.

section of the two graphs is the break-even point, Q, on the horizontal axis being the break-even quantity, S, on the vertical axis, the break-even money sales. Figure 1 shows how the problem is usually depicted in the literature.

Curvilinear approach. Figure 2 provides a diagrammatic demonstration of the curvilinear analysis. The graph of total cost, T, remains a straight line, as in 1, but the sales graph, S, becomes a curve, concave downwards as it rises to the right. Here, there are two break-even points, namely B_1 for a relatively high price, and B_2 for a relatively low price. Commercial undertakings are less interested in the break-even points in this kind of analysis than in the optimum price which the diagram indicates. The optimum price is the price that yields the maximum net profit, the measure of which is represented by the vertical distance between the cost graph and the sales curve where they are the furthest apart, *i.e.*, where they are parallel.

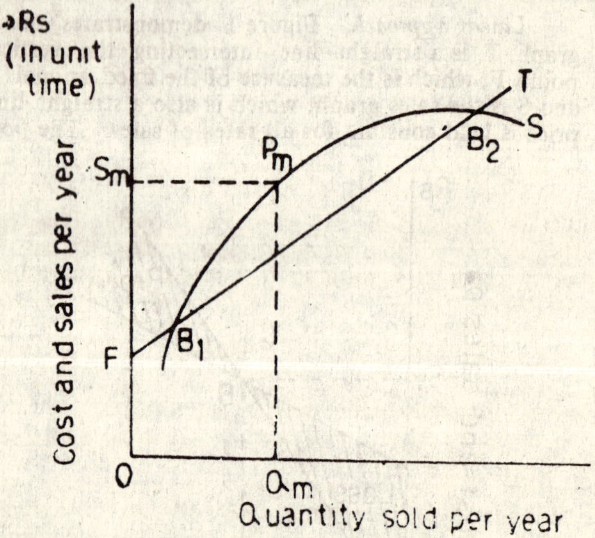

Fig. 2. Maximising net profit.

Applications. Profit planning, price fixing, make-or-buy decisions.

Break-Even Chart. A chart on which two curves are plotted ; the expected sales revenues from the production of various quantities of a product and the costs (fixed and variable) of producing those quantities. The curves intersect at the break-even point.

Brownian Motion Process. An additive stochastic process in a real variate x_t defined at time t such that $x_t - x_s$ has been Normally distributed with zero mean and variances $\sigma^2 \mid r-s \mid$, where σ^2 has been a constant.

Buffon's Needle. Refers to a statistically-based experiment for determining π. A piece of paper has been marked with parallel lines, distance a apart, and needle has been thrown onto the paper in a random manner (see Fig.). The probability that the needle crosses one of the lines has been put as follows :

$$\frac{2l}{a\pi}$$

where l is the length of the needle, and $l < a$. Thus if the needle is dropped a large number of times, it becomes possible to estimate π as

$$\frac{2l}{a} \times \frac{\text{Total number of throws}}{\text{Number of landings crossing a line}}$$

This experiment is a very slow way of finding π. After 10,000 throws, the first decimal place should be known with

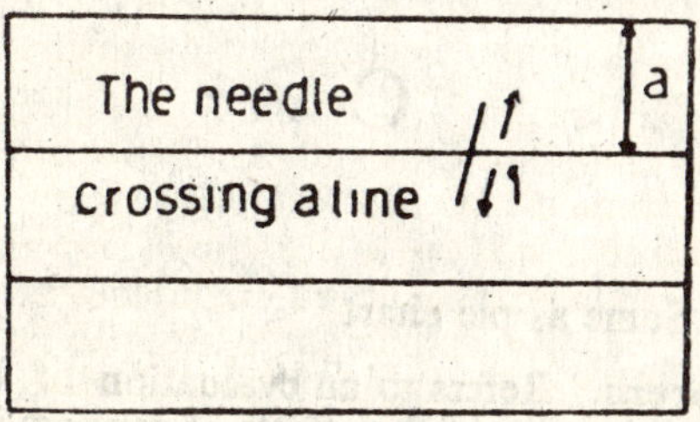

Buffon's needle.

reasonable confidence, the second not at all. It is also very difficult to ensure a random throw of the needle.

This experiment was originally intended as a method of calculating the stakes and winnings in the French gambling game france-carreau, rather than a method of estimating π.

Bulk Sampling. Refers to the sampling of materials which have been available in bulk form. That is to say, it has been the population which is in bulk ; the term does not mean the drawing of a sample in bulk. Examples of such sampling would be involving the sampling of a shipment of coal for ash-content, or of tobacco for moisture content.

Burke's Theorem. It is a theorem (1956) in queueing theory. This theorem states that if a single-server queueing process with exponential interarrival and service-time distributions is stationary, then it is evident that the departure process has been a homogeneous Poisson process with rate equal to the reciprocal of the mean interarrival time.

Burr's Distribution. It is a frequency distribution which was introduced by Burr (1942) where the cumulative distribution is having a simple algebraic form

$$F(x)=1-1/(1+n^c)^k \; n\geqslant 0 \; ; \; c, k\geqslant 1.$$

It has been one of a class wherein the distribution function, not the frequency function, has been defined in analytical terms.

Busy Period. In queueing theory it refers to a closed time interval such that all servers in a queueing system get occupied.

C

Cake Diagram. Same as pie chart.

Campbell's Theorem. Refers to an evaluation of the asymptotic distribution of the sum of the effects of random impulses which are acting with given intensity on a damped system.

The impulses are considered to occur at times according to a Poisson process with parameter λ. Each impulse is having a given intensity α and is having an effect $a\psi(t)$ after time t has passed. If $\theta(t)$ be the sum of the effects of all impulses eccurring prior to time t, then for $t \rightarrow \infty$ the mean and variance of $\theta(t)$ will be respectively.

$$\alpha \int_0^\infty \psi(t)dt, \; \lambda a^2 \int_0^\infty \{\psi(t)\}^2 dt.$$

Canonical Matrix. In general, it refers to any matrix which reduces to a canonical form ; *e.g.*, in the analysis of canonical correlations. It has been sometimes assigned specialised meaning.

Cantor-Type Distribution. It is a thoroughly pathological type of probability distribution for which there exists distribution function F and is continuous but for which there exists no density function.

Capon Test. It is a normal scores test for a difference between the scale parameters of two otherwise identical populations. The ranks R_i, $i=1, 2, \ldots m$ of the members of a random sample $X_1, X_2, \ldots X_m$ from one of the populations, after getting combined with a random sample $Y_1, Y_2, \ldots Y_n$ from the other, get transformed to scores which may be defined as the expectations of the squares of the order statistics of rank R_i in a random sample of $m+n$ realisations from a unit Normal distribution ; the sum of these scores has been the test statistic.

Capture/Release Sampling. It is a method of sampling which has been found to be specially suited to the estimation of the size of total populations of wild animals. It is also called capture/recapture sampling. The method was mainly introduced and used by Lincoln (1930). This method involves capturing, marking and releasing a random sample, say, of animals of a particular kind. Subsequently, a further random sample is considered and the proportion of marked animals in this sample will form the basis of estimates of total populations. [See also Lincoln Index.]

Cardinal Number. It is a number which is denoting quantity but not order in a group. Thus 1, 2, 3, ... have been cardinal numbers. The word cardinal has been generally used in contrast to *ordinal* which expresses position in a sequence, *e.g.*, first, second, third etc. (Also, see ordinal scale).

Cardioid Distribution. It is a circular distribution for which the density function is having a cardioid curve as its polar representation. It was given by Jeffreys (1948), and is having the density function :

$$f(\theta)=(2\pi)^{-1}\,[1+2_{p\cos}\,(\theta-\theta_0)],$$
$$0<\theta\leqslant 2\pi,\ |\,p\,|<\tfrac{1}{2}.$$

For $p=0$, it gets reduced to the circular uniform distribution.

Carleman's Criterion. There are certain conditions under which it is possible for two different distributions to have the same set of moments. Criteria therefore become desirable to decide when a set of moments will be able to determine a distribution uniquely. One such, given by Carleman (1925), states that a set of moments $\mu_1, \mu_2 \ldots$, not necessarily about the mean, will be able to determine a distribution uniquely if :

$$\sum_{j=0}^{\infty} \frac{1}{(\mu_{2j})^{1/2j}} \text{ (for distributions ranging from} -\infty \text{ to } \infty)$$

or

$$\sum_{j=0}^{\infty} \frac{1}{(\mu_j)^{1/2j}} \text{ (for distributions ranging from 0 to } \infty)$$

diverges. Cramer and Wold (1936) generalised this to the multivariate case.

Carli's Index. It refers to a simple index number of prices. It was proposed by Carli in 1764. Let the prices of a set of commodities in the base and given period are respectively p_o ; p_o', p_o'', ..., and p_n, p_n', p_n'', ..., Then, Carli's index number will be given by

$$I_{on} = \frac{1}{k} \Sigma \left(\frac{p_n}{p_o} \right)$$

where k denotes the number of commodities. The summation extends over all commoditions. It is therefore an unweighted index of price-relatives.

Carrier Variable. It is a name assigned to some numerical quantity for which the different values furnish the levels of the corresponding factor in a factorial experiment.

Cartesian Graph. See graph, abscissa.

Cartogram. It is a device which is used for displaying statistical information of a descriptive nature by using a symbol on a map. The symbolism may get various forms in accordance to taste, *e.g.*, dots or circles of varying density, or shading in black and white, or use of a full range of colours. The various forms of cartogram have been found to be especially convenient for portraying data according to geographical distribution.

Categorical Distribution. A distribution is considered to be categorical if the data could be sorted into categories in accordance to some qualitative description rather than by a numerical variable.

Categorical Scale or Nomiral Scale. It refers to a scale which sorts variables in accordance to category. For example, the cars in a park can be sorted according to make, the fish in a pond according to type, etc. For a categorical scale it becomes essential that the categories be exhaustive, exclusive and clearly defined. I ence every variable should belong to one and only one category, and there should be no doubt as to which one.

There have been two types of categories, a priori and constructed. *A priori categories* already exist, for example, a person's sex or year of birth ; *constructed categories* have been decided upon by research workers. A bad choice of categories can prejudice the outcome of an investigation. Categories may get indexed by numbers which do not correspond to any true order, such as the numbers on footballer's shirts.

Category. It is a homogeneous class or group of a population of objects or measurements. It is possible to style the category after one of the finite characteristics of the population or according to the limits of measurement for which observations are to be allocated to that category of frequency group. For example, it is possible to categorise people according to Sex (Male or Female) or Age next birthday (1—5 years ; 6—10 ; 15—15 ; 16—20 ; and so on).

Cauchy Distribution. It is the name generally assigned to denote the continuous distribution

$$dF=\frac{dx}{\pi(1+x^2)}, -\infty \leqslant x \leqslant \infty$$

or some simple transform to another origin or scale. The distribution is having no finite moments other, perhaps, than the arithmetic mean, which itself may be regarded to exist only by convention.

It has been a particular case of Student's distribution having one degree of freedom.

Causal Distribution. It is an alternative designation for deterministic distribution. As it is neither causal nor a distribution except in a degenerate way, it should be avoided.

Causal Relationship. Statistical analysis can go a long way towards establishing the existence of a causal relationship between two measurable factors. It cannot establish the nature of any causal relationship, nor can it be used for proving that any two factors are not causally related.

So far as the nature is concerned, statistical analysis may show that factors A and B are causally related ; it cannot show that A causes B or that B causes A, or indeed, that either one causes the other, since the relationship shown to exist between A and B may be the effect of factor C, *i.e.*, it may be that A and B are joint effects of C.

As to factors that, on the evidence of the statistics, seem not to be causally related, B and C may be joint causes of A, but C may behave so erratically relative to B that the statistics of A and B fail to reveal any causal relationship. Table contains a hypothetical example of a problem in three variables, in which A is dependent on B and C, and in which the statistics for A are built up from the equation $A=4B+10C$.

Table Causal relationship with three variables.

A	*B*	*C*
158	2	15
148	5	12
120	10	8
134	11	9
132	8	10

A comparison by inspection of the two series representing A and B, or for that matter, A and C, suggests little causal relationship between the two factors. Yet if we could compare A with B and C together, we would find the statistical relationship to be 100%. There is a statistical measure of the relationship between two or more statistical series : it is called the coefficient of correlation.

An understanding of the logic of causal relationships is of use in statistical research in all fields.

Causation. It is the production of an effect by a cause. It is not possible to prove. Causation statistically, although it may be very strongly suggested.

A common misunderstanding is concerned with the relationship between carrelation and causation. If there happens to be a strong correlation between the variables measuring the occurrence of two events A and B, it may be that

A causes B, $A \Rightarrow B$ or B causes A, $B \Rightarrow A$
or they both cause each other, $A \Leftarrow\Rightarrow B$

but it might also be that a third event C causes both A and B,

A
↗
C
↘
B

or that a more complex set of interactions is going on. It may also be that the correlation observed has been merely due to chance fluctuations in the variables involved and that there has been no causal reason for it at all. The statement 'Correlation does not imply causation' explains this situation.

The fact that there is a strong (negative) correlation between the mean number of cars per family and the mean number of children per family for the various countries of the world does not mean that owning a car does not allow you having children, or that having children stops you having a car. The situation has been more complicated when the affluence of the country has been involved; in affluent countries, birth rates have been low and car ownership is common.

Low birth rate
↗
Affluence
↘
High car ownership

(It could, however, be argued that affluence and low birth rates both cause each other).

Cause Variable. If a relation such as $y=f(x)$ has been interpreted in a causal sense, *e.g.*, y is has been considered as 'caused by' x, the latter is sometimes called a cause variable and the former an effect variable. The cause variable is also termed as an explanatory variable. See also regression and explanatory variable.

Cell Frequency. If the classification of a frequency distribution is done into categories, univariate or multivariate, the subcategories are sometimes termed as cells; the frequency with which observations come into a particular cell is termed as the cell frequency.

Censoring. We consider a sample to be censored when certain values are not known (or deliberately ignored) although their existence is known. The term is generally used in those cases where, of a number n of values existing, only the k smallest or k largest are observed or k might be expressed as a proportion of n. This is termed as Type II censoring; Type I is when the sample gets censored by reference to a fixed variate value. Whereas Type I censoring can be 'right' or 'left' censored it is more usual for Type II censoring to be only on the right of the variable. The ignoring of the existence of additional values or where such cannot exist is termed as truncation and refers to the population.

Census. The term used for the complete enumeration of a population or groups at a point in time with respect to well-defined characteristics : for example, Population, Production, Traffic on particular roads. In some cases the term is associated with the data collected rather than the extent of the collection so that the term Sample Census possesses a distinct meaning.

The partial enumeration which gets resulted from a failure to cover the whole population, as distinct from a designed sample enquiry, could be referred to as an 'incomplete census'.

In general, consus refers to an official periodic count of a population, including such information as sex, age, occupation. In India census is held every ten years. Heads of house holds are required to provide information about all those living under their roofs. The questions asked cover name, age family status, education and employment.

Centile. It is an abbreviated form of 'percentile'. It is not in general use, but frequently used in the statistical literature of psychological and educational testing.

Central Confidence Interval. We consider a confidence interval for a parameter θ with lower and upper limits t_1 and t_2 to be central when.

$$\Pr\{(\theta-t_1)<0\}=\Pr\{(t_2-\theta)<0\}.$$

The values t_1 and t_2 are then, in a sense, symmetrically placed with respect to θ.

Central Factorial Moments. In a strict sense, this expression means the factorial moments about the centre or some central value. Actually the term is sometimes applicable to the factorial moments which are calculated about some point near to a central value as distinct from the ends of the range.

Central Limit Theorem. A rule assuring that the sampling distribution of the mean approaches normal as the sample size increases, regardless of the shape of the population distribution from which the sample is selected.

This theorem states, taken if samples of size n are taken from a parent population with mean μ and standard deviation σ, then the distribution of their means will be approximately normal with

$$\textit{Mean}=\mu$$

$$\textit{Standard deviation}=\frac{\sigma}{\sqrt{n}}$$

As the sample size n increases, this distribution approaches the normal distribution with increasing accuracy (see sampling distribution). Hence in the limit, as $n \to \infty$, the distribution of the sample means→Normal, mean μ, standard deviation $\sigma/\sqrt{n}$.

If the parent population has been itself normal, the distribution of the sample means will be normal, whatever the sample size. If the parent population has been of finite size N, two possibilities arise :

(*a*) If the sampling has been done with replacement, the theorem stands as stated ;

(*b*) If there has been no replacement, the standard deviation of the sample means would be

$$\frac{\sigma}{\sqrt{n}} \sqrt{\frac{N-n}{N-1}}$$

The factor $\sqrt{\frac{N-n}{N-1}}$ has been termed as the finite population correction.

The central limit theorem forms the basis for much of sampling theory ; it can be summed up, as follows. If n is not small, the sampling distribution of the means is approximately normal, has mean$=\mu$ (the parent mean), and has standard deviation$=\sigma/\sqrt{n}$ (where σ denotes the parent standard deviation).

This theorem, although due to Laplace, was first proved rigorously by Liapounov (1901).

Central Tendency. Refers to the tendency of quantitative data to cluster around some variate value. The position of the central value is generally determined by one of the measures of location like the mean, median or mode. The closeness with which values cluster round the central value is measured by one of the measures of dispersion like the mean deviation or standard deviation.

Centre (of a Range). In a specialised sense the phrase centre of a sample' sometimes denotes the variate value which is midway between the two extreme variate values ; correspondingly 'centre' is sometimes used to refer to the mid-point of the range of a distribution.

Certainty. Refers to the condition of an event whose probability of occurrence has been equal to 1. It has been the decision environment in which only one state of nature exists.

Chain. It refers to a sequence of terms in such a way that each term is depending in some defined way upon the previous term or terms in the series ; for example, the chain-relative used in the calculation of index numbers upon the chain-base method.

The term chain has been also used in connection with stochastic processes where the value at one point has been determined by values at previous points apart from a random element ; or more exactly, the probability distribution at any point, conditional on certain previous values, has been otherwise independent of past history. The most common case has been the Markov chain.

Chain-Base Index Number. A type of index number, which changes its base (=100) and its pattern of weights from one period or from one data to the next. Both Laspeyres's formula and Pausche's formula can easily be converted to the chain base. Laspeyre's standard fixed-base price index number in its more practicable form is

$$P_{n\cdot 0}=\frac{\Sigma\left(q_0 p_0 \frac{pn}{p_0}\right)}{\Sigma(q_0 p_0)} \qquad (1)$$

where P is the price index, Σ the sign of summation, q the quantity, p the price, and subscripts n and 0, represent thg given and base periods, respectively, so that q_0, p_0 are the quantity and price in the base periods, and $P_{n\cdot 0}$ is the price index in period n, based on period 0=1.00. The correspondine chain-base formula is

$$P_{n\cdot 0}=P_{(n-1)\cdot 0}\,\frac{\Sigma\left(q_{n-1}\; P_{n-1}\,\frac{pn}{P_{n-1}}\right)}{\Sigma(q_{n-1}\; p_{n-1})} \qquad (2)$$

which gives the index number to the original base=1.00. It will be seen that the weights used are those of the previous period.

Chain Index. Refers to an index number in which the value at any given period gets related to a base in the previous period, as distinct from one which gets related to a fixed base. The comparison of non-adjacent periods is generally made by multiplying consecutive values of the index numbers, which, as it were, form a chain from one period to another. For example, if the value of the index for period 2 based on period 1 has been I_{12} and that for period 1 on period 0 has been I_{01} the chain indrx for period 2 based on period 0 is $I_{01}\times I_{12}$ (divided by 100 if the index numbers are based on 100 as the standard).

In practice chain index numbers are generally formed from weighted average of link-relatives, namely the values of magnitudes for a given period divided by the corresponding values in the previous period.

Champernowne Distributions. Refers to a three-parameter system of distributions which was suggested by Champernowne (1952) in connection with the distribution of incomes :

$$F(x)=1-(\cos\theta)\tan^{-1}\left[\sin\theta\Big/\left\{\cos\theta+\left(\frac{x_1}{m}\right)^{\alpha}\right\}\right]$$

where m denotes the median of x ; θ and α have been constants.

Chance 1. Refers to the unknown and unpredictable element that causes an event to have one outcome rather than another. 2. Refers to that which is described by probability.

Chance Variation. Refers to variation in statistical observations due to the action of random, as distinct from systematic, factors.

Changeover Trial. Refers to an alternative name for an experiment or trial which is using a crossover design.

Charlier's Check. It is an accuracy ckeck when can be used while we calculate the mean and standard deviation. It employs the fact that for a set of numbers $x_1, x_2, \ldots x_n$ with frequencies $f_1, f_2, \ldots f_n$,

$$\sum_{i=1}^{n} f_i(x_i+1)^2=\sum_{i=1}^{n} f_i x_i^2+2\sum_{i=1}^{n} f_i x_i+\sum_{i=1}^{n} f_i$$

Chebyshev's Inequality. The theorem states that : If a probability distribution is having mean μ and standard deviation σ, then the probability that the value of a random variable with that distribution differs from μ by more than $k\sigma$ is less than $1/k^2$, *i.e.*,

$$\Pr\{\,|\,x-\mu\,|>k\sigma\}<\frac{1}{k^2}$$

Chebyshev's Ineorem. No matter what the shape of a distribution, at least 75 per cent of the values in the population will fall within 2 standard deviations of the mean, and at least 89 per cent will fall within 3 standard deviations.

Chi Distribution. Refers to the distribution of the positive square root of the statistic called the χ^2 or chi-squared.

Chi-squared (or χ^2) Distribution. It is the distribution, with n degrees of freedom, of $\chi^2=X_1^2+X_2^2+\ldots+X_n^2$, where X_1,

$X_2, \ldots X_n$ have been n independent normal random variables with mean 0 and variance 1. As the χ^2 distribution is different for each value of n (where $n=7$) complete tables would be very cumbersome; they has been therefore usually given only for critical values for certain significance levels (*e.g.*, 99, .95, .90, .10, .05, .02, .01).

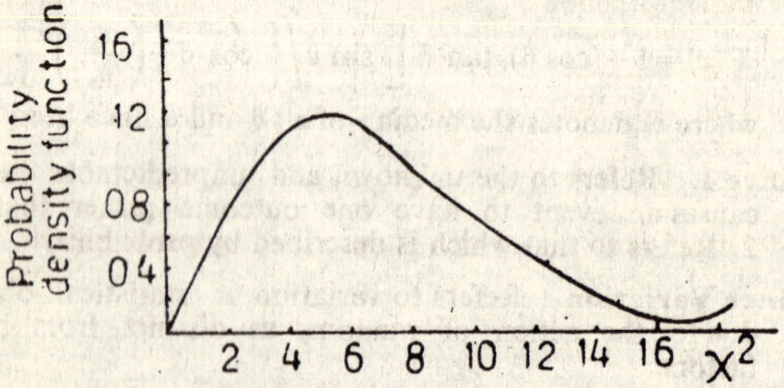

Chi-squared distribution. Graph for $n=7$.

The χ^2 distribution find use in a number of statistical tests including the chi-squared test, friedman's two-way analysis of variance by pank, and kruskal-wallis one-way analysis of variance.

This distribution was first given, apparently, by Asbbe (1863) and rediscovered by Helmert (1875) and K. Pearon (1900).

$$\chi^2 = \Sigma \frac{(f_o - f_e)^2}{f_e}$$

This formula says that the *chi-square statistic* (χ^2) is equal to the sum (Σ) we will get if we :

1. Subtract the expected frequencies, f_e, from the observed frequencies, f_o, for each category of our contingency table.
2. Square each of the differences.
3. Divide each squared difference by f_e.
4. Sum all the results of step 3.

Chi-squared Statistic. In a strict sense this expression should refer to a statistic which is distributed as chi-squared (X^2) namely as the sum of squares of independent standard Normal variates. For historical reasons, however, it more usually relates to a statistic of a particular kind which is, as a general rule, distributed more or less approximately in the χ^2 form. If a set of n values gets distributed over k classes such that the observed frequency in the jth class has been n_j and the theoretical (expected) frequency in that class has been v_j, the statistic.

$$\sum_{j=1}^{k} \frac{(n_j - v_j)^2}{v_j}$$

has been called the χ^2 statistic, or simply the value of χ^2, for the data. It has been widely used to test agreement between observation, as represented by the n_j, and hypothesis as represented by the v_j.

Chi-spuared (or χ^2) Test. It is a test of how well observed data fit a theoretical distribution. The data have been collected into groups, and the number of observations in each group denoted by O (observed). The expected number in each group E has been calculated from the theoretical distribution. The value of χ^2 thas been then worked out using

$$\chi^2 = \sum_{\substack{\text{All} \\ \text{groups}}} \frac{(O-E)^2}{E}$$

A small value of χ^2 implies the data fit the distribution well, a large value badly.

Chi-Squared Test. Occur in many ways, the most prominent being :

(*i*) An overall goodness-of-fit comparison of observed with hypothetical frequencies falling into specified classes ;

(*ii*) Comparison of an observed with a hypothetical variance in normal samples ;

(*iii*) Combination of probabilities from a number of tests of significance.

Chi-Squared Test for Variance. It is a test of the null hypothesis, that a particular sample is drawn from a parent population having a given variance. This test can only be applied if the parent population has been normal. It is dependent upon the fact that the statistic

$$\frac{n\sigma^2}{\sigma^2}$$

is having a χ^2 distribution with $n-1$ degrees of freedom, where n has been the sample size, σ has been the parent standard deviation estimated from the sample and σ has been the given parent standard deviation.

Chi-Statistic. It is equal to the square root of the more familiar chi-squared statistic.

Cigarette Card Distribution. If there have been k different types of cigarette cards being mixed equally frequently and the types are randomly distributed, the expected distribution of the k types in a collection of n cards ($n>k$) has been

$$\frac{k!}{(k-r)!r!}\ \frac{\Delta r(0^n)}{k^n}$$

where the second term is using the leading differences of powers of natural numbers.

Circular Chart. It is a method of diagrammatic representation whereby the components of a single total could be shown as sectors of a circle. The angles of the sectors will be prpor-tional to the components of the total. Additional visual aid could be obtained with coloured shading or cross-hatching. Also termed as a pie chart.

Circular Distribution. It is a frequency distribution of a variate which ranges from 0 to 2π, so that the frequency may be considered as distributed round the circumference of a circle. The term has been used especially of phenomena which have a period of 2π (by a suitable change of scale if necessary) so that the probability density at any point α would be the same as that at any point $\alpha+2\pi r$ for integral values of r. It has been generally expressed in terms of an angle θ. The analogue of the Normal distribution in this class, the so-called circular Normal distribution, would be given by

$$f(\theta)=e^{k\cos(\theta-\theta_o)}/2\pi I_o(k),\ 0\leqslant\theta\leqslant2\pi,$$

where I_o denotes a Bessel function of the first kind of imatgi nary argument. It was derived by von Mises in 1918. I is-possible to derive the expressions analagous to the standard deviation, variance, Pearson measure of skewness and kurtosis for the general circular distribution. For example, Mardia (1972) defines, for a set of values $\theta_1, \theta_2, \ldots \theta_n$. the mean direction θ_o as $\sum_{j=1}^{n} \theta_j/n$ and the circular variance as

$$S_o=1-\frac{1}{n}\sum_{j=1}^{n}\cos(\theta_j-\theta_0)$$

It could also be defined as $\Sigma(\theta_j-\theta_0)^2/n$ to which S_0 reduces for small $\theta'_j-\theta_0$.

The circular uniform distribution would be the simple form

$$f(\theta)=1/(2\pi),\ 0\leqslant\theta\leqslant 2\pi$$

Mardia also defines a circular triangular distribution, in symmetric form about $\theta=\pi$, as

$$f(\theta)=\frac{1}{8\pi}\{4-\pi^2p+2\pi p/\pi-\theta/\},\ 0\leqslant\theta\leqslant 2\pi,$$

$$p\leqslant 4/n^2,$$

and in asymmetric form as

$$f(\theta)=\frac{1}{2\pi}\{1+p\ (\theta-\pi)\},\ 0\leqslant\theta\leqslant 2\pi,$$

$$0\leqslant p\leqslant 1/\pi.$$

Circular Histogram. It refers to a form of diagrammatic representation for grouped angular data. In contrast to the linear histogram where the rectangular blocks representing frequencies have been erected on a horizontal base, the circular histogram is having the blocks attached to the appropriate intervals on the circumference of a circle.

Circular Lattice Distribution. A discrete distribution on the circle with

$$Pr\ (\theta=v+2\pi r/m)=p_r,\ r=0, 1, \ldots, m-1$$

and

$$p_r-0,\ \Sigma p_r=1.$$

Circular Mean Deviation. It is a measure of angular dispersion. For a sample of directions $\theta_1, \theta_2, \ldots, \theta_n$, the circular mean deviation about a central direction a would be defined to be

$$d_0=\pi-\frac{1}{n}\sum_{i=1}^{n}/\{\pi-/\theta_i-a\}$$

Similarly, the population circular mean deviation may be given as follows :

$$S_0=\pi-\xi\mid(\pi-\mid\theta-a\mid)\mid$$

which may be alternatively but as follows :

$$S=\int_0^{\pi}\theta dF\ (\theta+a)+\int_{\pi}^{2\pi}(2\pi-\theta)\ dF\ (\theta+a)$$

Circular Mean Difference. For a sample of n angular observations $\theta_1, \theta_2, \ldots, \theta_n$ it is measure of angular dispersion defined as

$$\overline{D}_0 = \frac{1}{n^2} \sum_{i=1}^{n} \sum_{i=1}^{n} \{\pi - |(\pi - |\theta_1 - \theta_j|)|\}.$$

For a population with distribution function F (θ), the circular mean difference would be correspondingly defined to be

$$\Delta_0 = \pi - \xi\{|\pi - |\theta_1 - \theta_2|)|\}$$

where θ_1 and θ_2 have been independently distributed as F (θ). This expression is having the alternative form

$$\Delta_0 = \pi - \int_0^{2\pi}\int_0^{2\pi} |\pi - \theta|\, dF(\xi)\, dF(\theta + \xi)$$

Circular Normal Distribution. It is an ambiguous term which is applicable to two entirely different distributions.

(1) A bivariate normal distribution in which $\sigma_1 = \sigma_2$ and $\rho = 0$, as the joint probability density function has been constant along circular contours.

(2) A distribution with frequency function

$$f_\theta(t) = [2\pi I_0(k)]^{-1} e^{k \cos(t - \theta_0)},$$

$$0 < t < 2\pi\ ;\ 0 < \theta_0 < 2\pi$$

where

$$I_0(k) = \sum_{j=0}^{a} (j!)^{-2} (\tfrac{1}{2}k)^{j2}$$

has been a Bessel function of zero order of imaginary argument. The distribution was derived by von Mises in 1918.

Circular Test. In constructing index numbers a decision has to be made as to the period upon which to base the index. If an index for period A based upon period B has been I_{BA} and for period B based upon period C has been I_{CB}, the circular test, derived by Irving Fisher, needs that the index for period A based upon period C, *i.e.*, I_{BA}, should remain the same as if it were compounded of two stages, the calculation of A on B and that of B on C. That is to say we should have

$$I_{CA} = I_{CB} I_{BA}.$$

A similar argument is applicable to comparisons between places. Few index numbers in current use have been found to satisfy this test.

Circular Triads. In paired comparisons which concern three objects X, Y and Z, if X has been preferred to Y, Y to Z and also Z to X the triad XYZ has been said (in the terminology of M.G. Kendall) to be circular. The circular triad exhibits inconsistent preferences, in the sense of that preferences, are not being made consistently on a linear scale, and no ranking will be possible.

Class. Refers to a collection or division of people, things, or values of a variable, sharing a common characteristic.

The world 'class' has some mild specialisation in the theory of frequency distributions. The total number of observations made upon a particular variate could be grouped into classes according to convenient divisions of the variate range in order to make subsequent analysis less laborious or for other reasons. A group so determined is termed as a class. The variate values which determine the upper and lower limits of a class are termed as class boundaries ; the interval between them has been the class interval ; and the frequency falling into the class has been the class frequency.

Class Boundaries. Refer to the boundaries for the classes when data has been grouped. For a continuous variable, the upper boundary of one class may be remain the same as the lower boundary for the next one. The reason for this has been that, in theory, if measured accurately enough, the variable will virtually always be one side or the other of the boundary. Hence the heights of adult men (in metres) could be grouped as :

1.65–1.70 1.70–1.75 1.75–1.80 etc.

In practice, however, measurements have been always rounded, in some way or other, and so it often becomes convenient to make it clear which class a variable apparently on the boundary has been to be assigned to. In that case the classes might be described as :

1.65–1.70–1.75–etc.

or $1.65 \leqslant\ < 1.70$ $1.70 \leqslant\ < 1.75$ $1.75 \leqslant\ < 1.80$

For a discrete variable, on the other hand, class boundaries could not be shared. The distribution of the annual egg yield per hen of a farmer's chickens might, for example, for grouped as

No. of eggs	0—49	50—99	100—149	150—199	200—249	250—299	300—349	Over 350
Frequency (no. of hens)	1	15	35	46	53	42	11	0

In this case the same figures do not appear in the boundares of two classes. There has been no need for this, as the number of eggs laid must be a whole number.

Class Frequency. The number of elements or individuals who fall in a particular class is called the class frequency. In the table below, the class frequencies are 108, 56, 55, 131.

	Age	*Frequency*
Class 1	0—19	108
Class 2	20—29	56
Class 3	30—39	55
Class 4	40 and over	131
Total		350

Classical Probability. The number of outcomes favorable to the occurrence of an event divided by the total number of possible outcomes.

Classification. For the purpose of convenient analysis, observations may be grouped into classes, so that the classes together exhaustively cover the entire range of values. The process of classification (*i.e* , the determining of class intervals, and consequently of class frequencies) is an arbitrary process. Classes are humanly determined, whereas attribute categories (*e.g.* sex) are usually self-evident.

In contrast with this general meaning of classification, the term classification statistic means a statistic used to determine from which population a sample originated, when it may have been drawn from one of a number of populations. In this context the term 'classification' assumes its more general non-statistical meaning of discrimination between one population and another.

Class Interval. The range of interval between the highest and lowest values allowed in a particular Class. For example, if we classify football terms by the number of points they obtained in the Football League last year, we might have a frequency distribution like that given below :

Number of points	*Frequency*
Fewer than 25	5
25—34	24
35—44	25
45—54	20
55 and over	18
Total	92

Note that two of the class intervals are open—*i.e.* in the lowest class no lower limit is defined (although 0 could be assumed), and in the highest class no upper limit is defined. The other three classes are closed and the interval is 10 points.

Class Limits. The smallest and largest values that go into any given class ; there are stated limits and real limits.

Class Mark. A term which is sometime used in elementary statistics, but obsolescent for advanced work, for denoting midvalue of the class interval.

The midpoint of a class—that is, its *class mark*—is calculated by averaging the lower and upper limits of that class.

$$\text{Class mark} = \frac{\text{Stated lower limit} + \text{Stated upper time}}{2}$$

When we are dealing with discrete variables and wide intervals, the midpoint of a class (its class mark) is calculated by using a slight modification of Equation (1). This enables us to work with more convenient values,

$$\text{Class mark} = \frac{\text{Stated lower limit} + \text{Stated lower limit of the next class}}{2}$$

Class Symbol. In the theory of attributes, it refers to a letter which denotes membership or non-membership of a class, *e.g.*, if A represents 'male' and α 'female', β represents 'living' and β 'dead', AB would represent 'living males'. In Yule's notation the symbol (AB) would represent the number of living males in the population under discussion.

Sometimes the symbol $\bar{B}$ has been used instead of β to denote not-B.

Classification Statistic. In general, it refers to a statistic which is calculated from a sample for the purpose of assigning the population from which the sample emanated to one of a number of classes. The term is some what synonymous with discriminant function.

Climograph. A graph which shows rainfall against temperature for a particular place. Points have been plotted for each month of the year, (mean values being taken), and then joined up to form a continuous 12-sided figure.

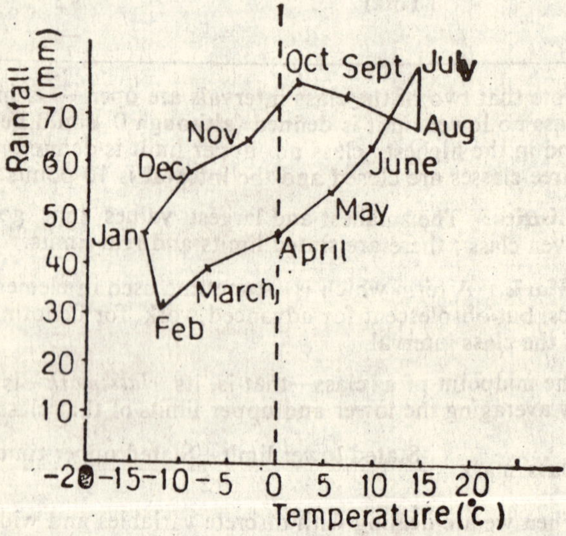

Climograph. Rainfall for Arkhangelsk (USSR)

Clisy. It is a term which was introduced by K. Pearson in connection with bivariate frequency arrays. For any fixed value of one variate x the distribution of y's is having a third mean-moment $\mu_3(x)$: the way in which $\mu_3(x)$ varies with x, or the corresponding $\mu_3(y)$ and y, expresses the variation in clisy of the distribution.

A curve which shows values of skewness for the different frequency arrays of one variate is termed as a clitic curve. In plotting these curves Pearson apparently used the measure of skewness $\beta_1(=\mu_3^2/\mu_2^3)$, Kendall (1943) employed the term in a slightly different sense to denote the graph of the third moment of the arrays against the corresponding variate value.

If all arrays are having the same skewness they are considered to be homoclitic ; if not, hetero-clitic. Pearson also defined a hetero-clitic system as nomic or anomic according as the skewness changes continuously or irregularly with the position of the array, but the terms are not into general use.

Closed Question. A question in which the respondent is given a limited number of responses from which to choose.

Closed Sequential Scheme. In sequential analysis the sampling generally continue until either an acceptance or a rejection boundary gets reached. The sample size has been not fixed but so as to avoid having (although perhaps, only rarely) to draw large samples before reaching a boundary it may be desirable to fix an upper limit to the sample size. The scheme is then termed as 'closed'. In the contrary case it is termed as 'open'.

Cluster. Refers to a group of contiguous elements of a statistical population. *e.g.*, a group of people who are living in a single house, a consecutive run of observations in an ordered series, or a set of adjacent plots in one part of a field.

Cluster Analysis. A generic term used to describe a large body of statistical procedures which are designed to group, or 'cluster', either elements or variables in a data set according to some well-defined characteristic. The objective of the analysis may be : (1) data reduction, *i.e.*, reducing the mass of data to manageable proportions. This is done : (*a*) by grouping similar elements together which will mean that it will be sufficient to consider the groups (of which there will only be a relatively small number) and not the elements (of which there may be a very large number) in subsequent analyses ; or (*b*) by forming clusters of variables (in particular attitudinal questions) so that summary scores for the respondents on the clusters of variables may be used instead of having to use all the answers to the individual questions. (2) To construct a typology or classification of rlements or variables-this may be appropriate in using sets of symptoms to construct a classification of illnesses for example. (3) To form classes or clusters of elements with particular characteristics—for example, identify deviant groups in a population. A form of cluster analysis has been used to identify the characteristics of drunked drivers. Such as analysis could be used for prediction purposes.

Cluster (Point) Process. Given a set of clusters which have been independently and identically distributed point processes, a cluster process consists of the superposition of all the clusters.

Cluster Sampling. Refers to a method of sampling in which the members of the sample have been chosen from one or several groups (or clusters) rather than at random. From a statistical point of view it is less satisfactory than simple random sampling, but it may well be more practical and/or economical.

Cluster Size. The number of elements in a cluster. In human populations, most clusters are of unequal size, *e.g.*, schools, firms, towns. However, in some cases clusters of equal size can be found, *e.g.*, the number of index cards in a file may be constant, or the number of dwelling units per block in a new housing development.

Cochran's Criterion. A criterion (Q) which was proposed by Cochran (1950) for the purpose of comparing percentage results in matched samples. In the general case the data get arranged in an $r \times c$ table with each row a matched group and each column a sample. The test criterion has been as follows.

$$Q = \frac{c(c-1)\Sigma(T_j - \bar{T})^2}{c(\Sigma u_i) - (\Sigma u_i^2)}$$

where T_j has been total success in jth sample (column) and u_i the total number of successes in ith matched group (row). Q is having a limiting distribution of χ^2 with $(c-1)$ degrees of freedom if the true probability of success has been the sample in all samples.

Cochran's Q The statistic used for cochran's criterion representing a adaptation of the chi-squared statistic test for significant differences between three or more sets or samples of matched observations.

Cochran's Q-Test. Refers to an alternative designation for Cochran's criterion in matched samples ; not to get confused with the coefficient of association due to Yule.

Cochran's Test. It is a test which was due to Cochran (1941), for homogeneity of a set of independent estimates of variance. It has been based on the ratio of the largest estimate of variance to the total of all the estimates.

Cochran's Theorem. A theorem on quadratic forms stated by Cochran in 1934. If $x_i (i=1, 2, \ldots, n)$ have been independent standardised Normal variates and $q_j = 1, 2, \ldots, k)$ have been quadratic forms in the variates x_i with ranks n_j $(j=1, 2, \ldots, k)$ and if $\sum_{j=1}^{k} q_j = \sum_{i=1}^{n} x_i^2$, then the necessary and sufficient condition for the q_j to be independent x^2 variates with n_j degrees of freedom respectively has been that $\sum_{i=1}^{k} n_j = n$.

Code. The term code denotes a number (a code number, or code) which is used to identify a category or class for categorized data. For example, in the frequency distribution below the numbers in the final column are the codes used to identify the classes in analysing the data.

Class	*Frequency*	*Code*
Under 15	78	1
15 and under 16	91	2
16 and under 17	65	3
17 and under 18	83	4
18 and over	103	5

Coded Data. Data which is translated from the form in which it was collected to one that is more convenient for analysis. In questionnaires, people are usually asked to reply by ticking appropriate boxes. When the replies have been analysed, each box usually is having a number (its code) associated with it, and it is these numbers which are analysed rather than the ticks themselves.

When figures have been grouped, it has been often possible to simplify calculations of mean and standard deviation by a particular kind of coding of the data.

Coding. Once the initial editing has been completed, the next operation is to prepare the data for analysis. In most surveys this involves coding the responses. The procedure of coding involves three stages. The first stage is to decide how to categorize the responses ; the second stage is to allocate numerical or symbolic values to the categories ; the third stage is to allocate each individual response to the appropriate category. The principal purpose of coding is to provide a convenient way of summarizing the responses (or observations) in surveys so that the data can be analysed more easily, generally on a computer. The set of categories for a question or set of questions, together with the numerical or symbolic codes to be allocated to them, is called a coding frame.

Coding may in effect be done before the fieldwork is carried out by printing the categories of response and the codes on the schedule itself. In this case all the interviewer (or in a self-completion questionnaire, the respondent) needs to do is to ring the appropriate code. Otherwise the coding must be done at a later stage.

Coding Error. Any errors which arise in the coding of the responses. The degree to which errors arise will depend on the ability of the coders, on the suitability and completeness of the coding frame and on the degree to which the editing has ensured the completeness and accuracy of the responses.

Coding Frame. The set of categories into which we classify the responses to a question or other stimulus, together with the numerical or symbolic codes we allocate to these categories, is called a coding frame. In many cases, the categories are self-evident and no serious problems arise.

Coding Sheet. The form on to which the codes for the responses to a survey schedule are written in preparation for punching. In some cases, where all the questions have pre-coded response, the punching may be done directly from the questionnaire or schedule.

Coefficient. 1. Refers to an index of measurement of characteristic. Examples have been the coefficient of alienation, and the coefficient of determination. 2. Refers to a numerical of constant factor in an algebraic expression. For example, 5 and 7 have been the coefficients of x^2 and x^3 respectively in the expression $5x^2+7x^3$.

Generally word coefficient has the same meaning as in mathematics, but occasionally it has been used to denotes a dimensionless statistic, *e.g.*, the moment-ratio β_2 as a coefficient of kurtosis or the coefficient of product-moment correlation. In this sense the word 'index' has been also used.

Coefficient of Alienation. Refers to a measure of the extent to which two random variables get unrelated. The coefficient of alienation is given as follows ;

$$\sqrt{(1-r^2)}$$

where r represents the product-moment correlation coefficient

Coefficient of Correlation. The square root of the coefficient of determination. Its sign indicates the direction of the relationship between two variables, direct or inverse.

Coefficient of Determination. Refers to the proportion of the variation of the dependent variable which is taken up by fitting in to the regression line. The variation in the set $y_1, y_2, ..., y_n$, with mean $\overline{y}$, is given as follows :

$$\sum_{r=1}^{n} (y_r - \overline{y})^2$$

It can be split into two parts :

$$\sum_{r=1}^{n} (y_r - \overline{Y})^2 = \underbrace{\sum_{r=1}^{n} (Y_r - \overline{Y}')}_{\text{Unexplained variation}} + \underbrace{\sum_{r=1}^{n} (Y_r' - \overline{Y})^2}_{\text{Explained variation}}$$

where (x_r, y_r') has been the point on the y on x regression line vertically above or below the point (x_r, y_r). Hence the unexplained variation has been the sum of the squares of the residuals $d_1, d_2, d_3...,$

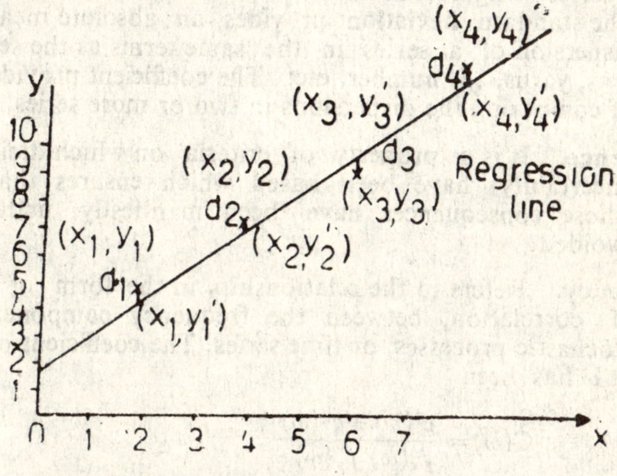

Coefficient of determination. Graph showing the residuals, d_1, d_2, d_3 and d_4 for the set of points $(x_1, y_1)...(x_3, y_4)$.

The coefficient of determination is thus Explained variation/Total variation, and has the value r^2, where r refers to the product-moment correlation coefficient. These statements have been also true if the word, variance, has been substituted for variation so that

$$\text{Coefficient of determination} = \frac{\text{Explained variance}}{\text{Total variance}} = r^2$$

Coefficient of Mean Deviation. An absolute measure of dispersion.

Coefficient of Multiple Correlation, R. The positive square root of R^2.

Coefficient of Multiple Determination, R^2. The fraction of the variation of the dependent variable that is explained by the regression. R^2 measures how well the multiple regression fits the data.

Coefficient of Variation. A relative measure of dispersion, comparable across distributions, which expresses the standard deviation as a percentage of the mean.

$$\text{Coefficient of variation} = \frac{\text{Standard deviation}}{\text{mean}}$$

Coefficient of variation is a standardized comparative measure of dispersion. It is equal to the standard deviation of a series of figures, divided by the arithmetic mean of the series. The standard deviation provides an absolute measure of the dispersion of a series in the same terms as the series itself—tons, yards, $$, number, etc. The coefficient provides a means of comparing the dispersions in two or more series.

Coherence. It is a property of criteria on which dicisions under uncertainty have been based which ensures that decisions whose consequences have been manifestly undesirable are avoided.

Coherency. Refers to the relationship, in the form of a measure of correlation, between the frequency components of two stochastic processes, or time series. The coefficient of coherence at ω has been

$$C(\omega) = \frac{c^2(\omega) + q^2(\omega)}{f_x(\omega)\, f_y(\omega)}$$

where $c(\omega)$ denotes the co-spectrum and $q(\omega)$ the quadrature spectrum and $f_x(\omega)$, $f_y(\omega)$ have been the spectral densities of the series x and y. The concept has been analogous to the square of the correlation coefficient (r). The functions in the expression for $C(\omega)$ are always obeying the 'coherence-inequality'

$$c^2(\omega) + q^2(\omega) \leqslant f_x(\omega)\, f_y(\omega).$$

Collapsed Stratum Method. Refers to a method for estimating the variance of the mean of a stratified sample where the sample has been based upon two random selections from each of a number of strata of equal size and the number of non-respondents results in too few completed strata. Two or more strata get amalgamated (collapsed) to form one stratum.

Collectively Exhaustive Events. The list of events that represents all the possible outcomes of an experiment.

Column Percentage. The percentage of any particular column in a given cell. For example, if the column total is 500 and the particular cell value is 25, its column percentage is 5.

Column Total, The total of all elements in a column (*i.e.*,

$$\sum_{j=1}^{i} a_j$$

where $a_1, a_2, \ldots\ldots a_i$ are each of the elements of the column).

Combinations. Refers to the ways of selecting a subset of a set, where the order in which the elements get selected has been of no importance. Thus the possible combinations of three letters out of A, B, C, D and E have been 10 in all :

A B C	A B D	A B E
A C D	A C E	A D E
B C D	B C E	B D E
C D E		

The number of ways of selecting r objects from a total of n, all different, has been denoted by :

$$nC_r = \frac{n\,!}{r\,!\,(n-r)\,!}$$

nC_r also yields the coefficient of $x^{n-r}\ y^r$ in the binomial expansion of $(x+y)^n$ (see also factorial).

Combination of Tests. Refers to the combination of a number of probabilities obtained from tests on different groups of data, undertaken so as to assess the probability of the test as a whole. One such test has been based on the fact that if k tests provide probabilities $p_1 \ldots\ldots, p_k$ the statistic $-2 \sum_{i=1}^{n} \log p_i$ has been distributed as χ^2 with $2k$ degrees of freedom, provided that the variates giving rise to the p's are continuous. In the case of discontinuity various modifications are needed.

Combinatorial. A term which denotes the number of different combinations of size r units which can be drawn from a set of n units. This quantity is used in calculating the coefficients of the binomial expansion and in computing the probability of selection of a particular sample in simple random sampling without replacement. It is written either ^nC_r or $\binom{n}{r}$ where r is the smaller number and n is the larger group. The formula for the calculation of combinatorials is

$$\frac{n\,!}{(n-r)\,!\,r\,!}$$

which may be simplified to

$$\frac{n\times(n-1)-(n-2)\times\ldots(n-r+1)}{1\times 2+3\times\ldots r}$$

Combinatorial Test. It is a test of significance in which the sampling distribution of the test statistic has been obtained by the algebra of combinatorial analysis.

Common Factor. In factor analysis the factor(s) have been classified in accordance to the way in which they contribute to the variables under analysis. Any factor which appears in two or more variates is termed as a common factor. If the factor appears in all the variates it is termed as a general factor. If it is common to a group of variates it is termed as a group factor. A factor appearing in only one variate is considered to be specific.

Common Factor Space. In one geometrical representation of a multivariate situation the variation has been considered as taking place in a space of which each factor represents a dimension. When there have been, say, m common factors and s specific factors the whole factor space is one of $n=m+s$ dimensions. Of this the m dimensional sub-space has been the Common Factor Space.

Communicate. A description of the relationship between two states in a Markov chain where they are mutually accessible from each other, that is, if A affects B and B conversely affects A.

For example, in the entry branching process the state 'customers' affects 'accounts paid' (0.5) and 'accounts paid' also affects 'customers' (0.8). Irrespective of other consequent states of both, these two states affect each other, or communicate.

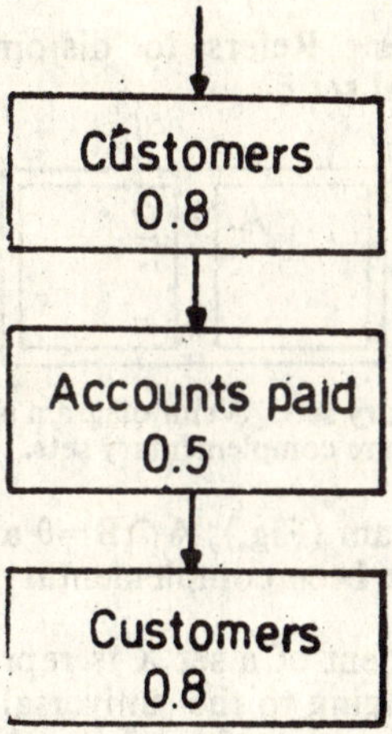

Comparative Mortality Figure. May be defined as the ratio of the standardised death rate to the crude death rate in a standard population. It may also be considered as an index number in the Laspeyres' form. [See also standardised mortality ratio].

Comparative Mortality Index. It is a variant of the comparative mortality figure. It has been a weighted average death rate, where the weights have been the mean of the actual (current) population and the standard population both expressed in proportions on a common basis of absolute size. In this sense it is termed as an index number of the Marshall-Edgeworth Bowley form.

Compensating Error. In general, it refers to any error which compensates for other errors. More specifically, a class of error having zero mean (unbiassed) and subject to the central limit effect, so that the occurrence of several errors will tend to cancel out and their effect get reduced as the errers cumulate. In this sense the term may not to be recommended.

Competition Process. It is stochastic process which is a two-dimensional extension of the birth and death process.

Complementary Events. Events which have been both exclusive and exhaustive. The events, 'The television is turned on' and 'The television is turned off,' have been considered as complementary. They cannot both happen at the same time (they are exclusive), and they cover all possibilities (they are exhaustive).

Complementary Sets. Refers to disjoint sets whose union has been the universal set ξ.

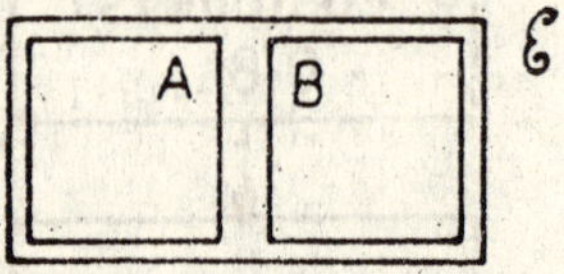

Complementary sets. Venn diagram showing that A and B are complementary sets.

In the venn diagram (Fig,), $A \cap B = \theta$ and $A \cup B = \xi$, so that the sets A and B have been complementary.

The complement of a set X is represented by X′, the set of all elements belonging to the universal set, but not included in X; X′ may be thought of as the set, not X. X and X′ have been complementary sets. Thus, $X \cap X' = \theta$, and $X \cup X' = \xi$.

Complete Class (of Tests). A class of tests has been complete if for every test outside the class there has been one within the class which has been uniformly better in the sense of a measurable decision function.

Complete Coverage. A survey or census can be described as complete if all of the elements in the population are included in the enumeration. Such a survey or census involves complete coverage of the population.

Complete Linkage/Farthest Neighbour. A form of cluster analysis. As in single linkage analysis the two most similar elements are first linked together. However, in order that another element be added to a cluster, the additional element must be sufficiently similar to all the elements already in the cluster. Thus the method will produce compact clusters, such that each of the elements in a cluster is closely related to (similar to) all the other elements in the cluster. The method is sometimes referred to as 'farthest neighbour' analysis since whether an element is included in a cluster depends on its similarity to the element already in the cluster which is most different (distant) from it.

Complete Regression. A concept which was introduced by Tukey (1958) in connection with estimation by order statistics where, if the distribution function $F(z/y)$ satisfies

$$F(z/y'') \leqslant F(x/y')\ y'' \leqslant y',$$

the regression has been complete negative ; complete positive regression being defined analogously.

Complete System of Equations. A term which is used mainly in econometrics for denoting the equations determining the behaviour of an economic system or part of such a system. The set of equations is considered to be complete when it includes all the determining equations governing the system, or a set from which all equations can be deduced. The point of emphasising completeness has been that in some cases it has been possible to estimate parameters taking place in an incomplete set of equations, but the estimators may then be biassed, a fact of which attention was drawn by Haavelmo in 1943.

Completely Randomised Design. Refers to a very simple form of experimental design in which the treatments get allocated to the experimental units purely on a chance basis.

Complex Demodulation.. Refers to a variant of the technique of demodulation where the basic series or stochastic process has been multiplied by a complex non-random function.

Complex Gaussian Distribution. A univariate complex Gaussian distribution has been based upon a random variable $z=x+iy$ where the real and imaginary parts have been each in a p variate Gaussian (Normal) form. This concept, introduced by Goodman, N.R. (1963) has been important in the spectrum analysis of multiple time series.

Complex Table. It is a table which shows the classification of a set of data is accordance to more than two different features : as distinct from the one or two features of the simple table. An example is that a human population might be tabulated in a complex table according to age, civilian status and sex. The complexity lies, not only in the manifold nature of the classification, but in the difficulty of printing the results in a convenient form. This method of tabulation is sometimes termed as multiple cross-classification.

Complex Unit. It is a statistical unit of record which is derived from a combination of two or more simple units. Examples are national income *per capita* or net-ton miles-per-train-mile (which is equivalent to the average train load) or pounds-per-week-earned-per-person.

Component Analysis. It may be defined as a branch of multivariate analysis which represents a k-dimensional variation as due to a number orthogonal components ; fewer than k if possible, but if not, in such a way that a few components are able to account for as much of the variation as possible. The components sought in practice have been linear functions of the original variates.

Component Bar Chart. It is a bar chart which shows the component parts of the aggregate represented by the total length of the bar. These component parts have been shown as sections of the bar with lengths in proportion to their relative size. Visual presentation could be aided by devices of cross-shading or colours.

Components of Variance. Refer to an alternative designation for Model II (or Second Kind) in analysis of variance.

Composed Poisson Distribution. A term which was introduced by Janossy *et al.* (1950) for describing the class of distributions now more commonly termed as Hermite distributions.

Composite Hypothesis. A composite statistical hypothesis may be frequently defined as a statistical hypothesis which is not simple. It is not entirely satisfactory and in practice the expression usually refers to a hypothesis which has been 'composed' of a group of simple hypothesis. For example, the hypothesis that a frequency function has been normal with unspecified mean and variance is composite since there has been a double infinity of values of mean and variance which, when specified, would give rise to a simple hypothesis.

Composite Index Number. It is a rather vaguely defined term which relates to an index number for which the component series have been from groups which are different in nature. The definition has been some what arbitrary in practice because much depends upon the point of view of both the complier and user of the index.

Composite Sampling Scheme. Refers to a scheme in which different parts of the sample have been drawn by different methods ; for example, a sample of a national population could be taken by some form of area sampling in rural districts and by a random or systematic method in urban districts.

Compound Frequency-Distribution. The expression is found in three senses : (1) If several sets of individuals, each with a frequency-distribution, get mingled to form a single set, the frequency-distribution of the latter may be considered to be compounded of the separate distributions ; *e.g.*, the distribution of heights of a population of human beings may get compounded of the distributions of heights of males and females.

(2) A frequency-distribution arising from the Convolution of variates may be considered to be compounded of the distributions of the individual variates. (3) If a distribution has been found to depend on a parameter θ which itself has a distribution, the distribution obtained by summing over θ has been said to be compound. This usage appears undesirable, because the word 'compound' then refers to the way in which the distribution was reached, not the properties of the distribution itself, So also, it appears preferable to refer to a distribution arrived at under (2) as 'convoluted'.

Compound Hypergeometric Distribution. It is a distribution which is produced by averaging the hypergeometric distribution for given x over all possible values of X according to a prior distribution. A number of well-known distributions may be regarded as special cases of the compound hypergeometric distribution. Examples are, the Polya, the binomial and the rectangular.

Compound Poisson Distribution. It is a name sometimes assigned to a distribution which gets resulted from a Poisson distribution of parameter λ where λ itself is having a distribution. If the distribution of λ's has been represented by $d\text{F}(\lambda)$ the probability of observing the number k

$$Pr\ (k) = \frac{1}{k\,!}\int_0^\infty e^{-\lambda}\lambda^k d^F(\lambda).$$

Computed *f*-Ratio. A statistic used to test the significance of the regression as a whole.

Computed *t*. A statistic used for testing the significance of an individual explanatory variable.

Concomitance (Concomitanza). Refers to the relation between two variates in time ; especially of the variation of two time-series in the same direction (positive concomitance) or in opposite directions (negative concomitance).

Concordance (Concordanza). Refers to a particular form of relationship (*connessione*) between quantitative variables, or qualitative variables when they are having comparable Modalities. If the variables have been such that positive values of one have been associated with negative values of another there has been said to be discordance (*discordanza*), that is to say, concordance tries to represent the sign as well as the intensity of the relationship.

Concordance, Coefficient of. A coefficient which particularly measures the extent of concordance, and is concerned with the extent of agreement among a number of ranked sets irrespective of paired comparison, unlik the coefficient of agreement, which is based on numbers of pairs of identical rankings.

The formula is

$$C = \frac{12\Sigma d^2}{m^2 n(n^2-1)}$$

where d is the deviation of the sums of rankings from their common mean $\left[\frac{m\,(n+1)}{2}\right]$, and where m is the number of rankings and n the number of objects.

Concurrent Deviation. Suppose (x_1, y_1), $(x_2, y_2), \ldots, (x_n, y_n)$ be pairs of observations taken from some convenient origin such as the means of the x's and y's. If x_i, y_i are having the same sign they are considered to exhibit concurrent deviation. A coefficient of correlation between x and y could be derived as the proportion of p of concurrent deviations to the total of n deviations. If the x's and y's have been distributed in a bivariate normal form an estimator of the correlation parameter has been given by sin $\frac{1}{2}\pi p$. [See also Kendall's tau].

Concurrent Validity. If a scale is constructed as an indicator of behaviour or of some independently observable criterion, its validity may be measured by the degree to which it succeeds in measuring this criterion. When the alternative measurement is carried out at about the same time, this is called concurrent validity. If the other measurement or observation takes place in the future, it is predictive validity.

Conditional. In some contexts this word may imply that there exist certain conditions which are obeyed by the quantities under discussion. It takes place most often in a specialised sense relating to variates. When a set of variates $x_1, x_2, \ldots, x_p, x_{p+1}, \ldots, x_q$ are having a joint frequency distribution the sub-distribution obtained by holding some of them fixed has been said to be conditional ; thus the distribution of $x_1, x_2, \ldots x_p$ for fixed $x_{p+1}, \ldots, x_q$, usually written

$$F(x_1, \ldots, x_p \mid x_{p+1}, \ldots, x_q)$$

has been the conditional distribution of $x_1, x_2, \ldots, x_p$ given $x_{p+1}, \ldots, x_q$. By extension, if certain functions of the x's, say statistics $t_1, \ldots, t_k$, have been held constant, the distribution of $x_1, x_2, \ldots, x_q$ has been conditional given $t_1, \ldots t_k$.

Conditional Failure Rate. An alternative term to hazard which finds use in life analysis of physical systems or components.

Conditional Probability. The probability p of an event occurring, given that another event has already occurred. The usual notation for the combined event of 'Event A occurring given that event B has occurred' is A | B (A given B). The probability may be given as follows :

$$p(A \mid B)=\frac{p(A\cap B)}{p(B)}$$

Example.: a card player cheats by looking at one of his opponent's cards. He sees a black picture card (King, Queen, or Jack) but he cannot make out which one. What will be the probability that the card has been a King ?
Event A : The card is a King.

Event B : The card is a black picture card.

$$p\ (B)=\frac{6}{52}\ p\ (A)=\frac{4}{25}\ p\ (A\cap B)=\frac{2}{52}$$

$$p\ (A \mid B)=\frac{p(A\cap B)}{p(B)}=\frac{2/52}{6/52}=\frac{1}{3}$$

Conditional probabiliry finds use in the definition of independent events. Events X and Y are independent if $p(X \mid Y)=p(X \mid Y')$ and $p(Y \mid X)=p(Y \mid X')$.

Conditional Probability Distribution. If the variates $x_1, x_2,\ldots, x_p$ have a joint probability distribution, the distribution of a subset of variates holding the others constant is called a conditional probability distribution. [See Conditional].

Conditional Regression. Refers to a regression which is estimated under certain conditions known *a priori* to apply to some of the parameters concerned ; for example, in estimating price and cross elasticities from time-series data the income elasticities involved can sometimes be considered to be known from cross-section data. It might be better to find an alternative name, because of the intimate connection between regression and conditional distributions.

Conditional Statistic. Refers to a statistic whose distribution has been conditional, that is to say, depends upon some quantity which is held constant ; the quantity in question has been generally itself some function of the variates entering into the statistic.

Conditional Test. A test of significance has been sometimes difficult to apply as the distribution of the test statistic involves unknown parameters of the parent population. This difficulty may sometimes be avoided by introducing restrictions on the sampling distribution, *e.g.*, by regarding only samples which have the same mean as that of the observed sample. It is equivalent to making the inference in a sub-population of samples, which have a fixed mean. The distribution and the inference based on it have been then said to be conditional.

Conditionally Unbiassed Estimator. An estimator t of a parameter θ is has been considered to be conditionally unbiassed with respect to statistics $u_1, \ldots, u_n$ if the expectation of t for constant $u_1, \ldots, U_n$ has been equal to θ. Symbolically

$$\xi(t \mid u_1, u_2, \ldots, U_n) = \theta.$$

Conditioning Event. An event which is known to have been taken place. The conditional probability of event A occurring, given that event B has occurred, is given as follows :

$$p(A \mid B) = \frac{p(A \cap B)}{p(B)}$$

In this, B has been the conditioning event.

Confidence Belt. Refers to the area between the upper and lower confidence limits.

Confidence Coefficient. Refers to the measure of probability a associated with a confidence interval which expresses the probability of the truth of a statement that the interval will include the parameter value.

Confidence Curves. A unified concept which was proposed by Birnbaum (1961) to include point, confidence limit and interval estimation.

Confidence Interval. Refers to that interval within which a parameter of a parent population gets calculated (on the basis of sample data) to have a stated probability of lying. The larger the sample size n, the smaller will be the confidence interval ; in other words, the more accurate will the estimate of the parent mean.

Confidence Level. The probability that we associate with an interval estimate of a population parameter indicating how confident we are that the interval estimate will include the population parameter.

Confidence Limits. Refer to the lower and upper boundaries of a confidence interval.

Confidence Region. If several parameters have been under estimate it is possible to define regions in the parameter space such that there will be assigned confidence α that the parameters are lying within them. It is the generalisation of the confidence interval to the case of more than one parameter and the domain so determined is termed as the confidence region.

Configuration. A set of n observations on a variate could be represented as a vector in n-dimensional space. A number k of vectors in the space may be considered as having geometrical properties of inter-relationship independently of the coordinate system. In factor analysis, the arrangement of variate vectors among themselves, and without regard to any frame of reference, is termed as their configuration.

The expression 'configuration of a sample' is also found in a somewhat different sense. The n observations define a point in the n-dimensional space and any set of such points lying in a subspace, *e.g.*, a hyperplane may be considered to have the same configuration.

'Configurational sampling' has been sometimes used as a synonym for grid sampling.

Conformity (Conformita). This term refers to the agreement between experimental results and those expected in accordance to some theoretical scheme.

Confounding. It is a device whereby, in large factorial experiments, the size of blocks gets limited by sacrificing some of the independent comparisons relating to the higher order interactions. These particular interactions may get deemed unimportant or of little practical consequence from a policy point of view.

The totality of possible treatment combinations does not get replicated in each block but gets divided amongst blocks in such a way that the main contrasts can be made within blocks but the others are not distinguishable from contrasts between blocks, with which they are regarded to be confounded.

Conjugate Distribution. If $F(x)$ has been a probability distribution function for which

$$M(t)=\int_{-\infty}^{\infty} e^{tx}dF(x)$$

has been finite in an interval l which contains the origin $t=0$, then for any value of t in this interval, the function

$$G(x, t)=M(t)^{-1}\int_{-\infty}^{x} e^{tu}dF(u)$$

has been a probability distribution function which has been said to be 'conjugate' to F(X).

Conjugate Latin Squares. Two Latin squares are said to be conjugate if the rows of one are the columns of the other.

Conjugate Ranking. Given two rankings of *n* objects, if one has been arranged in the natural order and the other (correspondingly rearranged) designated by A ; and then the latter gets arranged in the natural order and the first (correspondingly rearranged) designated by B ; then A and B are considered to be conjugate. For example two rankings of six objects $O_1, O_2, \ldots, O_6$:

	O_1	O_2	O_3	O_4	O_5	O_6	
	4	1	3	6	5	2	
	2	4	1	5	6	3	
rearranged as	O_2	O_6	O_3	O_1	O_5	O_4	
	1	2	3	4	5	6	
	4	3	1	2	6	5	A
and	O_3	O_1	O_6	O_2	O_4	O_5	
	3	4	2	1	6	5	B
	1	2	3	4	5	6	

yield the two rankings which are indicated by A and B as conjugate.

Conservative Process. It is a stochastic process which governs the behaviour of a population which has constant total size but the members of which can assume independently one of a finite number of states, the variation having transfer from one state to another.

Conservative Confidence Interval. A confidence interval or region is considered as 'conservative' when the actual confidence coefficient gets exceeded the nominal or stated value.

Consistency Checks. A part of the editing process in preparing survey data for analysis is to check whether the answers given to different questions are consistent with one another. One of the most common forms the checks take is to ask for the same information in two or more different ways. If, for example, at the beginning of the interview the respondent is asked for his date of birth, he might be asked at the end to give his age. The two answers should agree. Similarly if age, age at marriage and number of years of marriage are all obtained during the interview, then the number of years of marriage should be equal to age less age at marriage. The difficulty is that if the checking is done after the interview is completed and a discrepancy appears, there is no way of telling which of the answers is correct. However, the checks are valuable in providing an indication of the quality of the information collected.

Consistent Estimator. It is an estimator which gets converged in probability, as the sample size increases, to the parameter of which it is an estimator. An example of an inconsistent estimator has been the sample mean as estimator of θ in the distribution

$$dF = \frac{dx}{\pi\{1+(x-\theta)^2\}}, \quad -\infty \leqslant x \leqslant \infty.$$

Consistent Test. A test of a hypothesis has been consistent with respect to a particular alternative hypothesis if the power of the test tends to unity as the sample size tends to infinity ; and, similarly, it has been consistent with respect to a class of alternatives if it has been consistent with respect to each member of the class.

Constant. A mathematical term for a figure that remains constant or is assumed by some hypothesis to remain constant. There are the universal mathematical constants such as *pi* and *epsilon*, conversion factors such as knots to feet per second, and the so-called physical constants such as the velocity of light. Special constants of interest to management statisticians include marginal costs the elasticities of demand with respect to price and to other factors and other regression coefficients.

Constraint. A constraint in a set of data has been a limitation which has been imposed by external conditions, *e.g.*, that a number of variate values shall have zero mean, or that the sum of frequencies in a set of classes shall be a prescribed constant.

There has been another sense in which statistical data may be said to be constrained. It is the case of subnormal dispersion which was discussed by Lexis.

Consumer Price Index. A price index which is designed to measure changes in the cost of some specified standard of living.

Contagious Distribution. Refers to a class of probability distribution of a compound kind, generally derived from probability distributions dependent on parameters by regarding those parameters as themselves having probability distributions. The name is derived from the use of such compound distributions in the study of contagious events like accidents, occurrences of disease or 'persistence' in weather. [See also compound frequency distributions].

Contaminated Distribution. Refers to those distributions containing observations which have been of doubtful accurrancy or validity.

Contingency. It refers to the difference in the cells of the contingency table between the actual frequency and the expected frequency on the assumption that the two characteristics have been independent in the probabilistic sense. If f_{ij} denotes the frequency in the ith row and jth column and $f_{i\cdot}, f_{\cdot j}$ denote the respective row and column totals, and if the total frequency has been n, the difference in question will be

$$f_{ij} - \frac{f_{i\cdot} f_{\cdot j}}{n}$$

The square contingency is given as follows :

$$\chi^2 = \sum_{i.j} \frac{n(f_{ij} - f_{i\cdot} f_{\cdot j}/n)^2}{f_{i\cdot} f_{\cdot j}}$$

The mean-square contingency, usually denoted by ϕ^2, is given as follows :

$$\phi^2 = \chi^2/n.$$

Contingency Coefficient. A coefficient which attempts to measure the degree of association or dependence between two variates as expressed in a contingency table.

Contingency Table. A table having R rows and C columns. Each row corresponds to a level of one variable ; each column, to a level of another variable. Entries in the body of the tables are the frequencies with which each variable combination occurred.

Continuity. A parameter or a variate is considered to be continuous when it may take values in a continuous range. A frequency or probability distribution is sometimes considered to be continuous when it relates to a continuous variate, and sometimes when the function itself is continuous. Although in statistical practice the two have been aften equivalent it has been better to keeep the usage clear by referring to a distribution of a continuous variateor a continuous distribution of a variate as the case may can be.

It is possible to represent a continuous distribution by a discontinuous one, by dividing its frequency into categories based on interval measurements. Thus, for example, an age distribution, which is theoretically continuous, is frequency represented in terms of the frequencies in five-year age groups.

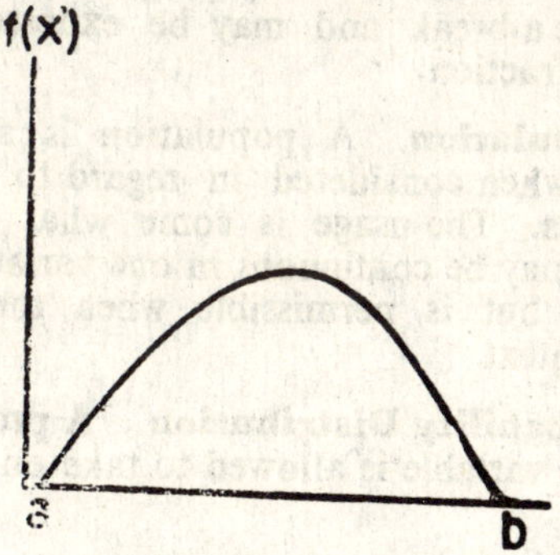

Continuous distribution

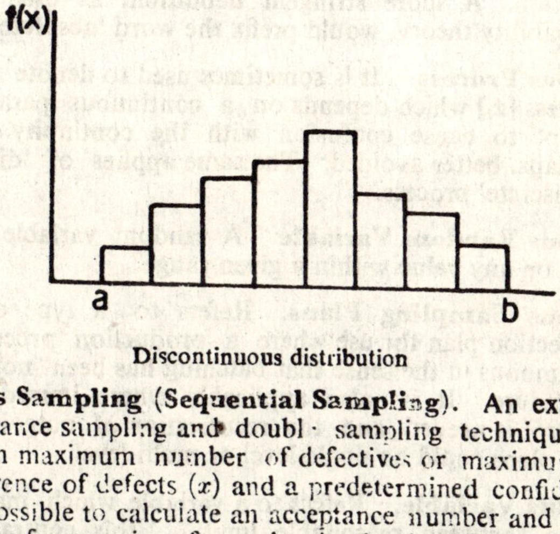

Discontinuous distribution

Continuous Sampling (Sequential Sampling). An extension of acceptance sampling and double sampling techniques. Using a given maximum number of defectives or maximum rate of occurrence of defects (x) and a predetermined confidence level, it is possible to calculate an acceptance number and a rejection number for any size of sample. As the size of sample increases, so will the acceptance and rejection numbers. If at any time the number of defects is less than the acceptance number, the sample is accepted. If the number of defects is greater than the rejection number, the sample is rejected. Otherwise sampling is continued.

Continuity Corrections. Refers to the corrections that may need to be applied when the distribution of a discrete random variable has been approximated by that of a continuous random variable.

Continuous Data. Data that may progress from one class to the next without a break and may be expossed by either whole numbers of fraction.

Continuous Population. A population is sometimes said to be continuous when considered in regard to some variate which is continuous. The usage is some what not very exact, as a population may be continuous in one variate and discontinuous in another ; but is permissible when the meaning is evident from the context.

Continuous Probability Distribution. A probability distribution in which the variable is allowed to take on any value within a given range.

Continuous Probability Law. A probability law is said to be continuous if it corresponds to a continuous distribution function. A more stringent definition, as used in advanced probability theory, would prefix the word 'absolutely'.

Continuous Process. It is sometimes used to denote a stochastic process $[x_t]$ which depends on a continuous parameter t.. It is apt to cause confusion with the continuity of x and is, perhaps, better avoided. The same applies of 'discontinuous' or discrete' process.

Continuous Random Variable. A random variable allowed to take on any value within a given range.

Continuous Sampling Plans. Refers to a type of sampling inspection plan for use where a production process has been continuous in the sense that batching has been not a rational procedure. It is also applicable where immediate process control is needed and the rectification of outgoing products. The plans could be single-level or multi-level.

Continuous Variable. Refers to a variable which may take any value (between reasonable limits) ; this contrasts with a discrete variable, which may take only certain values. Hence the height of an adult woman has been a continuous variable,

Continuous Variation. Items expressed in terms *e.g.*, of weight, volume, length and money value, are continuous in their variation. Items expressed in terms of whole numbers are not continuous and are said to be discrete in their variation.

Quantity figures of oranges sold by weight are continuous, those of oranges sold by number of units or creates are discrete.

Contrast. A contrast among the parameters in analysis of variance, or among the treatment and interaction effects in an experiment, has been a linear function of these quantities with known constant coefficients which sum to zero. It is considered to be an elementary contrast if it is having only two non-zero coefficient $+1$ and -1.

Control. There happen to be two principal ways in which this term finds use in statistics. If a process produces a set of data under what are essentially the same conditions and the internal variations are found to be random, then the process is considered to be statistically under control. The separate observations have been, in fact, equivalent to random drawings from a population distributed according to some fixed probability law.

The second usage is concerned experimentation for the testing of a new method, process or factor against an accepted standard. That part of the test which involves the standard of comparison is termed as the control.

Control Charts. (In quality control) diagrams which are used to give warning of when a machine or process is not functioning well. In this context manufacturing processes could be divided into two types. There are those whose product is either good or faulty, like electric light bulbs, and others where the quality of the product is determined by measurement. If a machine for making 5 cm nails is producing them with a mean length of 4.8 cm, such nails are unsatisfactory, as they do not meet the specification ; they can, however, still be used as nails, whereas faulty light bulbs are quite worthless.

In the situation where the product has been either good or faulty, an inspection scheme would involve recording the numbers of defective in samples of a certain size. This information could then be plotted on a control chart. This chart would be having two lines drawn on it, the inner control (or warning) limit and the outer control (or action) limit ; see Fig. The warning limit is usually set so that if the process is running correctly it gets exceeded by 5% of the samples, the action limit by 0.1%.

In cases where a measurement gets involved, two things require to be watched, the mean of the variable and its dispersion. A change in either could give rise to an unduly high

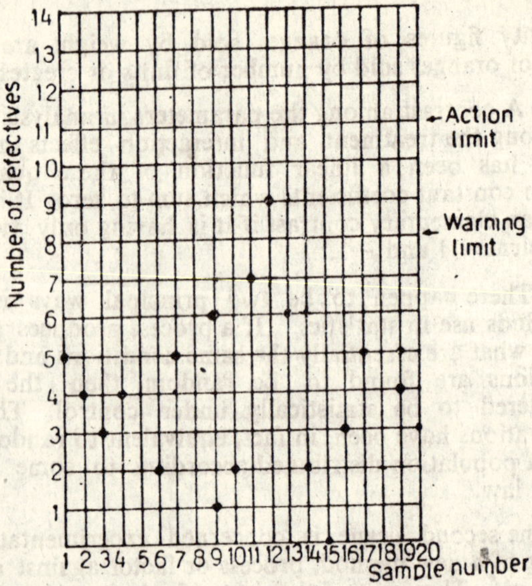

Control charts. The chart for numbers of defectives in samples of size 100 (process average, 40%).

proportion of the products being outside acceptable tolerance limits. Two control charts are thus required, one for the mean, the other for the standard deviation of the samples. The control limits have been typically set so that 5% of items from a correctly functioning machine lie outside the warning limits, 0.1% outside the action limits. If the standard deviation has not been known it has been often estimated using the tables on page 257, for converting range into standard deviation.

Control charts may also at times be used for measurements from individual items, rather than for the means and dispersions of the samples.

Control Experiment. An experiment which was designed to check or correct the results of another experiment by removing one of the variables operating in that other experiment.

Control Group. In experimental testing a group of persons or objects used as a standard of comparison or accepted norm with which to evaluate others among which a new process or method, or set of processes and methods, has been implemented.

Controlled Process. A production process subjected to valid quality control procedures so that the size, input content or output requirements of a given product are statistically stable. The means and variances of any or all of these variables are consequently predetermined. Any variation in size, input content or output requirements of the product will be the result of random effects.

Control Limits. Limits used in quality control for prescribing tolerances or allowable deviations from a required standard size in measurement or quantity in input and output. A given process may indicate that it is possible to manufacture a match with a length of 4 cm and standard deviation σ of 1 mm.

Control of Substrata. A term which is used in sampling inquires to denote the employment of prior knowledge of the population cell values in an n-way table formed according to the n factors which have been used in a scheme of multiple stratification.

Controlled Process. An industrial process is considered to be controlled when the mean and variability of the product remain stable. The variation has been then due to random effects or the combination of small factors of a non-cumulative kind.

Convolution. If $F_1(x)$, $F_2(x)$, ..., $F_n(x)$ be a sequence of distributions functions, then the distribution

$$F(x)=\int_{-\infty}^{\infty} dF_1(x_1)\ldots \int_{-\infty}^{\infty} F_n(x-x_1-\ldots-x_{n-1})\, dF_{n-1}(x_{n-1})$$

is termed as the convolution of the distributions. The relationship has been sometimes written as follows :

$$F(x)=F_1(x)*F_2(x)*\ldots F_n(x).$$

If the associated variates have been independent, $F(x)$ has been the distribution function of their sum.

Corrected Moment. Refers to a moment of a set of observations which is adjusted for some effect such as the bias arising from its being calculated from a grouped frequency distribution rather than from the original data.

Correction for Continuity. If a statistic is essentially discontinuous, but its distribution function is being represented approximately by a continuous function, the probability levels could sometimes be more accurately determined by entering the tables of the continuous function, not with the actual values of the statistic, but with slightly corrected values. These values are then considered to be corrected for continuity.

Correction for Grouping. When data have been grouped into frequency distributions the approximation which becomes necessary by reason of having to regard frequencies as being concentrated at the mid-points of class intervals may be able to give a bias to the calculations of the moments of the distribution. Under certain conditions it becomes possible to correct for this effect, the best-known corrections being due to Sheppard.

Correlation. Means interdependence between two or more random variables. If two variables have been such that, when one has changed the other will also change in a related manner, they are considered to be correlated. Variables which are independent are not correlated. Correlated variables are sometimes termed as related variables.

Correlation Analysis. A statistical technique used to discover and quantity relationships between variables. The two techniques of correlation and regression analysis are closely related. Two variables are correlated if a change in one variable is associated with a change in another. This need not imply a causal relationship since the movement of both variables may be influenced or controlled by other variables which may or may not have been included in the experiment or the data. Correlation can be positive (when a positive change in one variable is accompanied by a positive change in the other), or negative (when a positive change in one variable is accompanied by a negative change in the other). The degree of correlation between two variables is measured by a statistical parameter called the covariance.

Correlation Coefficient. Refers to a measure of association between two random variables. If two random variables have been such that when one gets changed the other will do so in a related manner, they are regarded to be correlated (see correlation). Variables which are independent are not correlated. The correlation coefficient has been a number between —1 and 1. It could be calculated from a number of pairs of observations which are normally referred to as points (x, y). A coefficient of 1 implies perfect positive correlation, —1 perfect negative correlation, and 0 no correlation. See also alienation.

Correlation Index. An absolescent term which refers to a quantity which purported to measure correlation where regression relationships between the variates concerned had been not linear.

Correlation Matrix. For a set of variates $x_1,\ldots,x_n$ with correlations between x_i and x_j denoted by r_{ij}, the correlation matrix has been the square matrix of values (r_{ij}). Its determinant is the correlation determinant. The matrix is symmetric since $r_{ij}=r_{ji}$.

Unless otherwise specified the diagonal elements r_{ij} $(i=1,\ldots,n)$ have been unity and in psychological work the matrix with unit diagonals has been said to be complete.

Correlation Ratio. In a bivariate frequency table with variates x and y the correlation ratio of x on y may be defined by.

$$\eta^2_{xy}=\frac{\Sigma(\bar{x}_i-\bar{x})^2}{\Sigma(x-\bar{x})^2}$$

where the summation in the numerator occurs over y-arrays, $\bar{x}_i$ is the mean of the ith array and $\bar{x}$ is the mean of x in the whole distribution ; and the summation in the denominator takes place over all the values. It may be considered as the ratio of the variance between arrays to the total variance.

Correlation Surface. Refers to an almost obsolete synonym for a bivariate frequency surface.

Correlation Table. Refers to the frequency-table of a bivariate distribution. The difference between a correlation table and a contingency table has been that the former generally represents a grouped frequency distribution with intervals defined in terms of the variate values and therefore having a natural order and clearly defined width.

Cost Function. In sampling theory, it refers to a function giving the cost of obtaining the sample as a function of the relevant factors affecting cost. It could be able to relate to only a part of the entire cost, *e.g.*, by providing for the cost of collecting the sample but not for the cost of tabulation.

Cost is a measure of input. Whether input is a function of output, or output a function of input, is the kind of question that runs all through management statistics. When the method of least squares is applied, it becomes a problem as it does with mathematical model building.

Cost-of-Living Index. An index number of retail prices made by the Government of and published monthly. Cost-of-living-index is the old name ; it is now called the *retail-price-index* or the *index of retail prices.* As the index is used extensively for a wide range of purposes the Department goes to a great deal of trouble in compiling it. The formula used is Laspeyres's index. Its pattern of weights based on family-budget inquiries, which are conducted from time to time. It has been said that this makes the index more of the Paasche type than of the Laspeyres type, but this is not so. A Paasche index is currently weighted ; the retail-price index is never currently weighted : the pattern of base-period weights is merely brought up to date from time to time, a process that makes it what might be called a hybrid chain-base index. The present index includes rent, rates, other housing costs, public transport, travel, other retail services, fuel and light, meals out and holidays, as well as food, clothing, durably household goods and other goods.

Applications. As a deflator to convert money values to real values and in wages negotiations.

Countable Set. It is a non-empty set, each of whose members could be labelled with a distinct positive integer. If the elements of such a set could be labelled using the integers 1, 2, 3,...N, the set is considered to be a *finite set*, with N elements. If no such N exists, the set is considered to be *countably infinite*.

The set of prime factors of the number 462 has been finite, with 4 elements (2, 3, 7 and 11).

The set of all prime numbers has been countably infinite.
The set of all points on a plane has been infinite, but not countably infinite.

The set of all points on a plane has been infinite, but uncountable.

Counting Distribution. It is a probability distribution which is formed by the number of events taking place in a fixed period of time. If the counting period starts just after an event the distribution may be known as synchronous distribution (Haight, 1965).

Covariance. The first product moment of two variables about their means. The formula for the calculation of the covariance is

$$\frac{1}{N}\,\Sigma(X_i-\bar{X})(Y_i-\bar{Y}) \quad \text{or} \quad \frac{1}{N}\left(\Sigma X_i Y_i-\frac{\Sigma X_i \Sigma Y_i}{N}\right)$$

where X_i and Y_i are corresponding values of each variable, and N ts the number of observations.

Covariance Analysis. It is an extension of the analysis of variance to cover the case where members falling into the classes bear the values of more than one variate. Interest centres on one of these (chosen as dependent variate) and the question has been whether its variation between classes has been due to class effects or to its dependence on the other variates which themselves vary among classes. This is discussed by considering the regression of the dependent variate on the other variates and the variation of the regressions (or equivalently of covariances) among classes. The technique has been found to be similar to that of variance analysis but considerable more complicated.

Covariation. Refers to the joint variation of two or more variates.

Coverage. A term which has been used in sampling in two senses : (1) It denotes the scope of the material collected from the sample members (as distinct from the extent of the survey, which refers to the number of units included) ; (2) This term is also used to mean the extent or area covered by the sampling as in expressions such as '50 per cent coverage', which implies that one-half of the population under discussion have been examined.

Cox's Theorem. A theorem given by to D.R. Cox (1952). This theorem states specific conditions under which a sequential test of a mean of a Normal distribution against a composite hypothesis could be constructed. The test has been based upon the application of the sequential probability test to transformed values of the original observations.

Craig Effect. When certain short-cut methods of periodgoram analysis are used on series consisting of a random element superposed upon a scheme of harmonic components there results a bias in the estimate of the periods. The magnitude has been found to depend upon the variance of the random component.

Criterion. This word finds use in statistics in its colloquial sense in a number of contexts, *e.g.*, the likelihood criterion for testing hypotheses.

In earlier literature the phrase '*the* criterion', otherwise unqualified, has been used to denote a function which distinguishes between various types of Pearson curves. The criterion is as follows :

$$k=\frac{\beta_1(\beta_2+3)^2}{4(2\beta_2-3\beta_1-6)\,(4\beta_2-3\beta_1)}$$

where β_1 and β_2 denote the Pearson measures of skewness and kurtosis.

Critical Region. For the purpose of testing a statistical hypthesis hypothesis at 1 per cent, 5 per cent or any other level of significance, the total sample space is divided into regions which are mutually exclusive. Sample statistics may either fall into the region of acceptance, which is usually the larger region, or into that of rejection, which is usually the smaller one. The region may consist of two areas. For example, in a two-tail test the region is a combination of two areas, one at each extreme of the curve. The critical region is that of rejection (*i.e.*, of non-acceptance). In the student's t-test, for

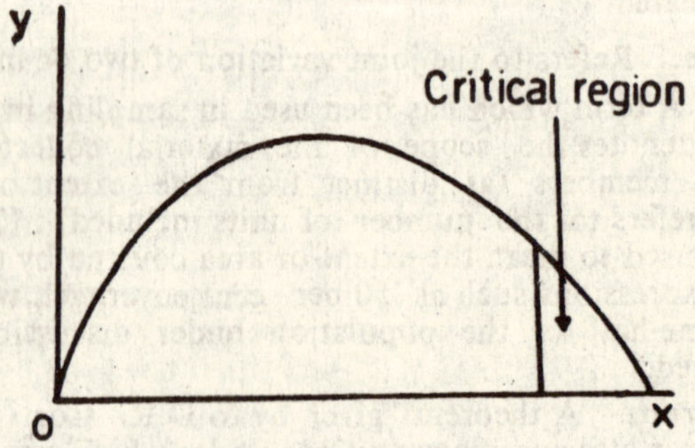

example, it is represented by the areas under the curve at one or both extremes beyond a particular critical value, but this is only one well-known example of a critical region. Generally the term may be applied to the rejection region for any tests of hypothesis.

Critical Value. Refers to the value of a statistic corresponding to a given significance level as determined from its sampling distribution ; *e.g.*, if $\Pr(t>t_0)=0.05$, t_0 has been the critical value of t at the 5 per cent level. The critical value for any test depends on the significance level at which the test is being applied, and on whether it is a one-or two-tail test.

Cross-Classification. One of the simples methods of multiple-regression analysis applicable to problems in up to five variables. It is, in practice, an adaptation of the simple-regression method of group averages to problems in three or more variables. Although a method of multiple-regression analysis in its own right, it appears to be used mainly in empirical model building for determining the type of mathematical model that best fits the statistics. Where the type of mathematical model is known to be linear, a two-tier system of cross-classification can be used for a regression analysis, but where the object of applying the method is to determine the appropriate type of model, a three-tier system is necessary.

Cross Intensity Function. In a renewal process events of several types might be taking place along the one dimensional time axis. The cross intensity function has been a method which was proposed by Cox & Lewis (1967) for dealing with the relationship between these different types of events considered jointly against time.

Cross-Section Series. A term applied to any statistical series that relates to different things or different places at the same time, as distinct from a time series which relates to the same thing or the same place at different times.

Sometimes hybrid series are used for analytical purposes. If, for instance, the transport firm had only five vehicles in use, then the three statistical series might relate to each of the vehicle's operations in each of three weeks, in which case the statistics could be described as hybrid.

Cross-Section Analysis. It is a stastical analysis of variations in data which are obtained from a set of sources at the same time. Compare time series analysis.

Cross-Tabulation. A classification in a two-way table of the elements of a population according to their scores or values on two variables is called a cross-tabulation. Each cell contains the number of elements which have common scores or values on each of two-variables. The table below presents the cross-tabulation of a group of fifty people by sex and age group.

Sex / Age group	*Under 20*	*20—39*	*40—59*	*60 and over*	*Total*
Male	4	10	8	3	25
Female	3	9	8	5	25
	—	—	—	—	—
Total	7	19	16	8	50

Such a table presents in summary the information on all fifty people for the two characteristics, age and sex. It is in effect a two-way frequency distribution.

Cross-Validation Criterion. This criterion involves the controlled or uncontrolled division of a data sample into two subsamples. The choice of a statistical predictor, including any necessary estimation, has been made on one subsample and then the assessment of its performance has been made by measuring its. predictions against the other subsample.

This procedure is also termed as the predictive sample re-use method.

Crossed Weight Index Number. An index number is regarded to have crossed weights if it is resulted from two subsidiary index numbers, with different weights, after the application of some process of averaging.

The most commonly quoted crossed weight formula has been that of Fisher's 'ideal' index which has been the result of geometrically crossing (averaging) the index number formulae attributed to Paasche and Laspeyres. It could be written as a price index :

$$I_{on}=\sqrt{\left(\frac{\Sigma p_n q_o}{\Sigma p_o q_o}\ \frac{\Sigma p_n q_o}{\Sigma p_n q_n}\right)}$$

where *o* and *n* denote subscripts relating to the base year and current year respectively. The Marshall-Edgeworth-Bowley index is also having crossed weights in this rense.

Cubodial Lattice Design. This experimental design, has been a development of the cubic lattice design in much the same way as the rectangular lattice has been a development of the square lattice. The cubic lattice design has been found to be suitable for numbers of treatments which are prefect cubes (k^3) whereas the cuboidal lattice design has been appropriate to a number of treatments of the form k^2 $(k+1)$.

Cumulant. The cumulants have been constants of a frequency distribution defined in terms of the moments by the identity in t

$$exp\left(\sum_{r=0}^{\infty}\frac{k_r tr}{r\,!}\right)=\sum_{r=0}^{\infty}\frac{\mu_r' tr}{r!}$$

They have been thus given by the coefficients in the expansion of a power series formed from the logarithm of the characteristic function of a variable, it such an expansion exists. The earlier name for the quantities had been semi-variant or half invariant, a term introduced by Thiele. The word cumulant has been due to Cornish and Fisher (1937).

Cumulative Distribution (Probability) Function. A synonym for the distribution function.

Cumulative Error. Refers to an error which, in the course of the cumulation of a set of observations, does not tend to zero. The relative magnitude of the error does not then decrease as the number of observations increases.

Cumulative Frequency. The frequency with which a variable is having a value less than or equal to a particular value.

Example : The yield of each tree in an orchard of 200 trees has been recorded, and the figures have been grouped as follows :

Yield (kg)	0—10	10—20	20—30	—30—40	—40—50	50—60	—60—70	70—80	80—90
Frequency (trees)	0	4	16	32	46	48	34	12	8

The corresponding cumulative frequency table is as follows :

Yield	⩽10	⩽20	⩽30	⩽40	⩽50	⩽60	⩽70	⩽80	⩽90
Ferequncy	0	4	20	52	98	146	180	192	200

It is derived from the frequency table by adding the frequencies from the left, or cumulating up to the entry in question. This can be drawn as a cumulative frequency graph or *ogive* (shown below). Such a graph allows the median, quartiles, precen-

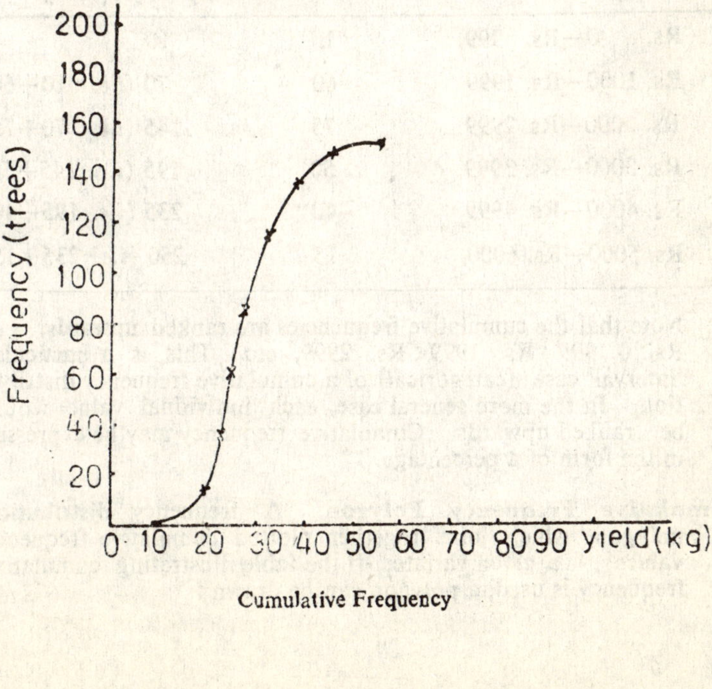

Cumulative Frequency

tiles, etc., to be read easily. It also allows questions like, 'Estimate how many trees yielded less than 43 kg', to be answered readily. If the points get joined with straight lines rather than a curve, the diagram is termed as a *cumulative frequency polygon.*

Cumulative Frequcncy (Probability) Curve. See distribution curve.

Cumulative Frequency (Probability) Function. A synonym for the distribution function.

Cumulative Frequency Distribution. A tabular display of data showing how many observations lie above or below certain values.

Cumulative Frequency. The sum of all the frequency values of a given variate (x) below and inclusive of a given measure of x for ordinal or interval level variates.

Example : The distribution of earnings in a particular office is shown in the following table :

Salary scale	*Frepuency (i.e. number of recipients)*	*Cumulative frequency*
Rs. 0—Rs. 999	10	10
Rs. 1000—Rs. 1999	60	70 (*i.e.* 10+60)
Rs. 2000—Rs. 2999	75	145 (*i.e.* 70+75)
Rs. 3000—Rs. 3999	50	195 (*i.e.* 145+50)
Rs. 4000—Rs. 4999	40	235 (*i.e.* 195+40)
Rs. 5000—Rs. 6000	15	250 (*i.e.* 235+15)

Note that the cumulative frequencies are ranked upwards, *i.e.*, Rs. 0 999 ; Rs. 1999 ; Rs. 2999, etc. This is a particular 'interval' case (categorical) of a cumulative frequency distribution. In the more general case, each individual value would be ranked upwards. Cumulative frequency may be expressed in the form of a percentage.

Cumulative Frequency Polygon. A frequency distribution polygon, which joins together plotted cumulative frequency values f a given variate. If the table illustrating cumulative frequency is used, a polygon can be drawn :

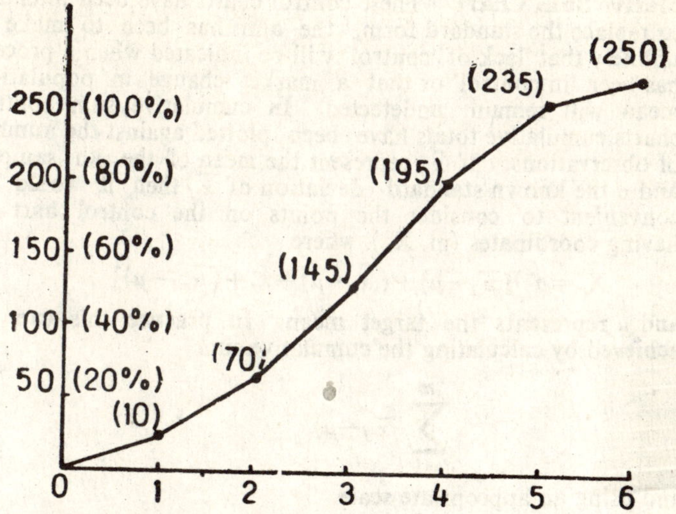

A cumulative frequency curve is a smooth continuous depiction of variate frequencies, but otherwise resembles the cumulative frequency polygon shown above.

Cumulative frequency may be expressed in the form of a precentage. For example, in the cumulative frequency polygon depicted above, the figures in brackets on the right-hand side of the vertical axis are cumulative percentage frequencies, which express cumulative frequencies as percentages of the total number of observations.

The advantage of cumulative percentage frequency curves and ogives is that two or more curves depicting distributions of unequal numbers of observations may be compared using the same diagram.

Cumulative Normal Distribution. Refers to the cumulative frequency function (distribution function) of the Normal distribution.

Cumulative Process. It refers to a development of the regenerative process, and a generalisation of the renewal process, introduced by Smith (1955). It deals with accumulation of regeneration points, or some attribute taking place at such as point, with the passage of time. An alternative formulation by D.G. Kendall (1948) presented this process as a variant of the standard birth and death process.

Cumulative Sum Chart. These control charts have been intended to replace the standard form, the aim has been to make it unlikely that 'lack of control' will be indicated when a process has been 'in control' or that a market change in population mean will remain undetected. In cumulative, sum control charts cumulative totals have been plotted against the number of observations. If $\bar{x}_j$ represent the mean of the jth sample, and σ the known standard deviation of $\bar{x}_j$ then it would be convenient to consider the points on the control chart as having coordinates (m, X_m) where

$$X_m = \sigma^{-1}[\bar{x}_1 - \mu) + (\bar{x}_2 - \mu) + \ldots + (\bar{x}_{ml} - \mu)]$$

and μ represents the target mean. In practice this can be achieved by calculating the cumulative sum

$$\sum_{j=1}^{m} \bar{x}_j - \mu)$$

and using an appropriate scale.

Cumulative Sum Distribution. Generally, if $x_1, \ldots, x_n$ are a set of variates, the distribution of $X_n = \sum_{i=1}^{n} x_i$ would be the cumulative sum distribution.

Curtate. A word which is used in vital statistics to denote the integral number of years, as distinct from the nearest number of years, for which a given state has existed. For example, if an assurance matures in 3 years 9 months, the curtate duration would be 3 years. An analysis which uses curate periods would, in most cases, be subject to downward bias.

Curve Fitting. An expression which is being used in two rather different senses in statistics : (*a*) In one sense, it is used to denote the fitting of a mathematically specified frequency curve to a frequency distribution (*cf.* Pearson curves) ; (*b*) In the second sense, it is used to denote the fitting of a mathematical curve to any statistical data capable of being plotted against a time or space variable, *e.g.*, regression data or time series.

Curvilinear Correlation. An expression which is used to denote correlation in bivariate data when the regressions are not linear. Strictly speaking it has been a misnomer, a correlation being a number and not a function admitting curvature, and has been perhaps better avoided.

Curvilinearity. A relationship between two or more variates which is depicted graphically by any locus other than a straight line. For example, in time series analysis the accelerating rate of increase in prices is called a curvilinear trend and represented by a steepening curve of prices plotted against time. Curvilinearity could represent a logarithmic, quadratic or higher order relationship, or a more complex function.

Curvilinear Regression. It is a type of regression which is not linear. A form often considered has been that for which the dependent variate has been expressed as polynomial in the independent variables.

Curvilinear Relationship. An association between two variables that is described by a curved line.

Curvilinear Trend. Refers to a trend which is not linear. It could be expressed as a polynomial, a more complicated mathematical expression such as a logistic curve, or by some smoothing process like a moving average.

Cut-Off. Refers to the artificial truncation of a sampling process at a point when it becomes apparent that enough data are being collected for the purpose in view.

Cyclical Movements/Fluctuations. In *time series analysis* cyclical movements are sequential, repetitive, but not necessarily regular, movements on either side of a given trend line. The object of time series analysis in to discover the extent to which a time sequence of variable values may be predicted by : (1) a trend line ; (2) cyclical movements ; (3) episodic movements caused by unusual events ; and (4) residual random fluctuations. Thus cyclical movements are oscillatory movements superimposed on the trend line in wave like fashion. In business statistics, for example, they may include both (*a*) season variations and (*b*) variations resulting from 'trade cycle' conditions, although the definition is usually reserved for (*b*) in some statistical texts and seasonal fluctuations are isolated separately.

In other branches of statistics, migratory cycles, political cycles (*e.g.* between Presidential elections in the United States) fall into the same category.

Cyclic Component. It is a non-seasonal component in a time series which varies in a recognizable cycle. An example is the frequency of sunsports, which has an 11.2-year cyclic component (Fig.).

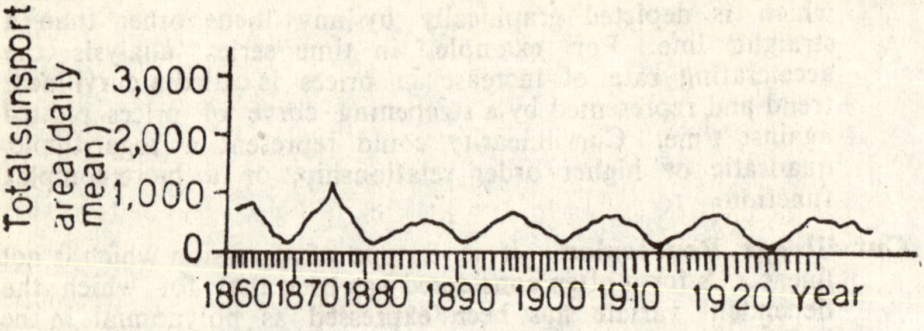

Cyclic component. The 11.2-year cycle of solar activity (after W.J.S. Lockyer).

Cyclic Design. Refers to a class of balanced and partially balanced experimental design in which the blocks have been constructed by cyclic permutations of the treatments.

Cyclic Order. Refers to an arrangement of n permutations of n objects such that if the first has been denoted by $1, \ldots, n$ the second is $2, \ldots, n, 1$ and the third $3, \ldots, n, 1, 2$, etc. The process offers some of the basic latin square designs ; for example; with four treatments

ABCD
BCDA
CDAB
DABC.

D

Daniels' Test. It is a distribution-free test for trend (1950), the basis of which has been to treat a time series

$$(t_i, X(t_i)) \quad i=1, 2, \ldots, n$$

as a sample of n observation from a bivariate population, and test for trend by means of Spearman's ρ.

Darmois-Koopman-Pitman Theorem. According to this theorem under certain regularity conditions on the probability density, a necessary and sufficient condition for the existence of a sufficient statistic of fixed dimension would be that the probability density belongs to the exponential family.

Data. The plural of the Latin word 'datum' (=given). The word can mean any information which is 'given' or provided for the solution of a problem. Because more than one unit of information is necessary for solving every problem, the word is plural. Data must be quantified and structured before being usable for the solution of a statistical problem. They are measured in statistical units, which must be : (1) relevant to the enquiry or problem ; (2) homogeneous (*i.e.*, of the same type) ; and (3) stable.

In accounting statistics a measure of homogeneity if often achieved for comparing dissimilar units, for example, the standard hour, a measure for work which a typical individual performs in an hour.

Data are often divided into two categories, primary and secondary indicating nearness to source or otherwise. For example, data collected by means of experiment, interview, observation, questionnaires or survey are primary, while items of data obtained by the researcher from a previous source or publication are secondary. Often the categories are unclear in practice.

Data Array. The arrangement of raw data by observations in either ascending or descending order.

Data Base. A store of a large amount of information which is specially in a form that could be handled by a computer. Information in a data base has been stored as a number of records or files, each of which usually has entries under a number of headings of fields.

Data Capture. Any process which is used for converting information into a form that could be handled by a computer.

Data Point. A single observation from a data set.

Data Set. A collection of data.

Datum Zero. A base level from which values of a variable have been taken. For example, heights of tides have been given with respect to a datum zero, called Low Water Ordinary Springs, (LW), and defined physically by a mark on Newlyn harbour wall ; dates are measured with respect to a datum zero, the start of year 1 AD ; longitude is measured with respect to the Greenwich meridian as datum zero.

In the assumed mean method of working out averages, the assumed mean has been a datum zero for the numbers involved.

David's Empty Cell Test. It is a distribution-free test. According to this test a continuous population possessing a particular completely specified probability density function against the general alternative hypothesis that it has not.

The idea behind the test has been that if the area under the hypothesised density function has been divided into a number of vertical strips of equal area, the bases of which have been regarded as cells, then a simple random sample from the actual population would tend to be more evenly dispersed among the cells if the hypothesis were true than if it were false. The number of empty cells should thus give rise to a reasonable index of the truth or falsity of the hypothesis.

D^2 Statistics. A statistic which was introduced by Mahalanobis (about 1930) and is regarded as a measure of the 'distance' between two populations with differing means but identical dispersion matrices. The distance between the populations

$$\triangle^2=\Sigma a_{jk}(\mu_{1j}-\mu_{2j})(\mu_{1k}-\mu_{2k})$$

is termed as Mahalanobis' generalised distance. The sample value of $\triangle^2$ would be as follows :

$$D^2=\sum_{j,k=1}^{p} a_{jk}(\bar{x}_{1j}-\bar{x}_{2j})(\bar{x}_{1k}-\bar{x}_{2k})$$

and (a_{ik}) would be the inverse of the pooled dispersion matrix. Developments of this concept, principally by the Indian school of statisticians, have been sometimes collectively termed as the 'distance method'.

Death Process. Refers to a stochastic process in continuous time defined on the state space of non-negative integers, representing the decay of a population in which only deaths may occur. The expression is generally confined to processes for which instantaneous transitions to the next lower integer only are possible, the probability of a transition from n to $n-1$ in time dt being asymptotically $\mu_n dt$; μ_n might be time-dependent.

Death Rate. It may be defined as the number of deaths in a given period divided by the population exposed to risk of death in that period. For human populations the period is generally one year and if the population is changing in size over the year the divisor is considered as the population at the mid-year. The death rate as so defined is termed as 'crude'. If some refinement has been introduced by relating mortality to the age and sex constitution (or other factors) for comparative purposes the rate will be considered to be standardised.

Decapitated Negative Binomial Distribution. It is a negative binomial distribution from which the zero class is missing. The term 'decapitated' was used by Yule to characterise a turncated distribution from which the zero class only is removed.

Decile. One of a series of partition values, known as quantiles, which divide the frequency of a group, set, sample or population into equal proportions. As their name implies, deciles divide the total frequency of a set of variate values into ten equal partition.

Decimal Places. Refers to the figures to the right of the decimal point that are needed to express the magnitude of a number to a specified degree of accuracy.

Decision Function. It is a rule of conduct which, at any stage of a sampling investigation, informs the statistician whether to take further observations or whether enough information has been collected, and in the latter case, what decision to make upon it. At each stage beyond the first the decision function has been a function of the preceding observations.

Until the development of sequential methods decision functions had been mostly of the simple type, based on a fixed sample size, which enjoined the acceptance or rejection of a hypothesis or set limits to a parameter under estimate. The above definition gives a sequential situtation wherein the investigator may not able to reach a decision about the hypothesis but proceeds to take further observations.

Decision Point. Branching point that requires a decision.

Decision Space. In sequential analysis and the theory of decision functions, the decision space has been the set of all possible decisions.

Decision Tree. A statistical device for estimating the optimum size of order for perishable goods, or of a plant for making any kind of standard goods. As originally depicted, the decision tree consists of a number of main branches, each representing a size of order or plant and each with one or more secondary branches representing judgements of the financial consequences. The main branches grow out of a box labelled D (for decision) situated on the left of the picture and extending to the right, the secondary branches extending further to the right, and the whole tree lying prone as though it had just been felled.

In its most logical and practicable form, the decision tree exploits the frequency distribution. The decision tree is not to be confused with the economic ordering quantity which is concerned with stores control through ordering frequency.

Decomposition. Refers to the act of splitting a time series into its constitutent parts by using statistical methods. A typical time series in often considered as composed of four parts :

(*a*) A long-term movement or trend ;

(*b*) Oscillations of more or less regular period and amplitude about this trend ;

(*c*) A seasonal component ;

(*d*) A random, or irregular, component.

Any particular series should not exhibit all three of these but those which are present are assumed to act in an additive fashion, *i.e.*, are superimposed ; and the process of determining them separately is one of decomposition.

A more modern approach (Wold, 1938) needs to decompose the series into deterministic and non-deterministic elements. This is termed as Wold's decomposition or predictive decomposition.

Decreasing Hazard Rate. Conventionally, a decreasing hazard rate may be defined as one which does not increase ; it can, therefore, be constant.

Deep Stratification. A term used to denote stratification by a substantial number of factors with respect to the marginal distribution of the factors only.

Defective Probability Distribution. A probability distribution usually assigns a probability measure over the appropriate interval in such a way that the sum of all the assigned probabilities would be unity. However, probability measures have been sometimes encountered which assign a total mass, say p, which has been less than unity. Such a measure is termed as a defective probability measure with defect $1-p$: the probability distribution has been a defective distribution.

Defective Sample. A sample which is incomplete or results from an incomplete survey or examination. Two examples of defective samples are ;

1. a sample taken from a telephone directory in an enquiry related to the study of wealthy people. Although it is valid to assume that most people whose incomes are above a given level have telephones, some of the most wealthy people are ex-directory, *i.e.*, their numbers do not appear in published directories. Such a sample could therefore be defective ;

2. a sample which is incomplete, because some records or completed questionnaires have been lost.

Defective Sample. It is a sample which has resulted from an inquiry which has been incompletely carried out, *e.g.*, because certain assigned individuals are not examined, because records are lost, or because (in plant or animal experiments) certain members are dead.

Defective Unit. In quality control, this term refers to a unit which does not reach some prescribed standard and therefore gets, rejected, in contrast to an effective unit.

De Finetti's Theorem. This theorem is due to de Finetti (1939). According to this theorem, a simple sample that generates a permutable binary sequence can always be considered as having been obtained by binomial sampling in which the type I probability P has a Type II initial distribution which is unique.

Defining Contrast (or Relation). In the analysis of half-replicate factorial designs the comparison between those treatment combinations which are used with those that are not used is termed as the defining contrast (Finney, 1946). It defines the halves into which the whole replicate gets divided.

Deflation of Values. An economic time series measured at current prices can be converted to a constant price series, giving a better indication of changes in volume by dividing it by an index of the prices of its components. This technique is referred to as deflating the series. Some difficulties involved are the choice of base period in terms of whose prices the deflated series is expressed and the fact that the relative importance of the components of the series will change as their relative prices change. This means that a suitable index of prices often cannot be constructed and a proxy measure must be used such as a retail or wholesale price index.

Degenerate Distribution. Refers to an alternative designation for deterministic distribution.

Degrees of Freedom. This term has been use in statistics in slightly different senses. It was introduced by Fisher on the analogy of the idea of degrees of freedom of a dynamical system, that is to say the number of independent co-ordinate values which have been necessary to determine it. In this sense the degrees of freedom of a set of observations (which *ex-hypothesi* are subject to sampling variation) has been the number of values which could be assigned arbitrarily within the specification of the system. For example, in a sample of constant size n grouped into k intervals there have been $k-1$ degrees of freedom because, if $k-1$ frequencies are specified, the other has been determined by the total size n; and in a contingency table of p rows and q columns with fixed marginal totals there are $(p-1)(q-1)$ degrees of freedom.

Degrees of Randomness. Strictly speaking there have been no degrees of randomness, except in the theory developed by M.G. Kendall under which randomness may be defined as relative to an assigned test domain. The expression is however (rather regrettably) introduced into the theory of stochastic processes.

A random function $\phi(t)$ possesses n degrees of randomness if n is the smallest integer such that if n values are known at distinct points of time then it is almost certain that a functional relationship exists between $\phi(t)$ $[T>t_n]$ and these available values.

Demand Schedule. A statistical schedule designed to show the effect of changes in price on the quantity of a brand of product sold in unit time, and on the sales proceeds.

Demodulation. A technique which was used in spectrum analysis of time series for the detecting and describing the more unusual kinds of non-stationary behaviour. The method involves multiplying a series by a suitable function and applying a Filter to leave only the frequency in a narrow bank.

Demography. The study of population size, composition, and distribution, and the patterns of change therein. The narrowest conception of demography views it as the study of vital statistics (the study of birth and death rates and related statistics).

De Morgan's Laws. The laws which state that, for sets or events $A_1, A_2, \ldots A_n$,

$$(a) \qquad \left(\bigcup_{i=1}^{n} A_i\right)' = \bigcap_{i=1}^{n} A_i'$$

$$(b) \qquad \left(\bigcap_{i=1}^{n} A\right) = \bigcup_{i=1}^{n} A_i'$$

These may be stated in words as :

(*a*) 'The complement of the union of events is the same as the intersection of their complement'.

(*b*) 'The complement of the intersection of events is the same as the union of their complements'.

These laws could be illustrated by this simple example from sets.

$\xi=\{a, b, c, d, e, f\}$, $A_1=\{a, b, c\}$ and $A_2=\{b, c, d\}$.

In this case, ***de Morgan's first law*** gives

$A_1 \cup A_2=\{a, b, c, d\}$, and so $(A_1 \cup A_2)'=\{e, f\}$;

$A_1'=\{d, e, f\}$, $A_2'=\{a, e, f\}$, and so $A_1' \cap A_2'=\{e, f\}$;

so $(A_1 \cup A_2)'=A_1' \cap A_2'$.

de Morgan's second law gives

$A_1 \cap A_2=\{b, c\}$, and so $(A_1 \cap A_2)'=\{a, d, e, f\}$.

$A_1' \cup A_2'=\{a, d, e, f\}$.

So $(A_1 \cap A_2)'=A_1' \cup A_2'$.

Density Function. See frequency function.

Departure Process. It is the stochastic process which is formed by the intervals between the departure times from the service counter or station of a queueing situation. This departure process has been of importance in the study of tandem queues (Jackson, 1957).

Dependent Samples. Samples drawn from two populations in such a way that the elements were not chosen independently of one another, in order to allow a more precise analysis or to control for some extraneous factors.

Dependent Variable. A variable which depends on another (the *independent*, or explanatory variable). If we draw a graph, the dependent variable is generally drawn on a vertical (y) axis, the independent variable on the horizontal (x) axis (Fig.)

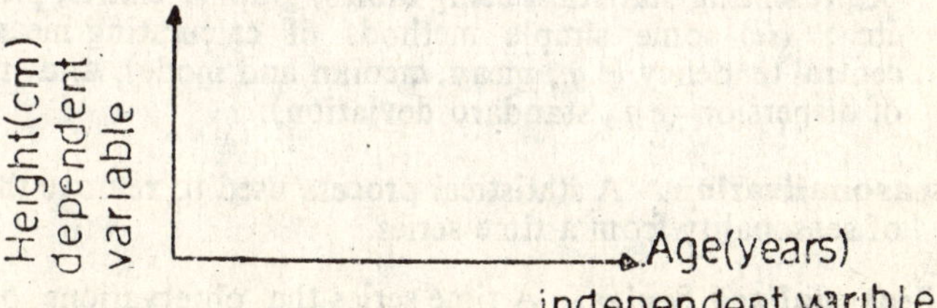

Dependent variable.

If the graph were drawn of a girl's height, measured every birthday, against her age, her height would be the dependent variable, her age the independent. Her height depends on her age but her age does not depend on her height, just on the length of time she has been alive.

Derived Statistics. This type of statistics is obtained by an arithmetical operation from the primary observations. In this sense, almost every statistic could be 'derived. The term is especially used to denote descriptive statistical quantities which are obtained from data and are primary in the sense of being mere summaries of observations, *e.g.*, population figures have been primary and so have been geographical areas but population-per-square-mile is a derived quantity.

Descriptive Indices (Indici Descrittivi). An index (or coefficient) is called descriptive if it shows a relationship concerning each value of the variable, as distinct from the global or mean index or coefficient (*indice global or medio*) which relates only to the aggregate or the mean of the values. Means, standard deviations, mean differences, etc., have been global indices ; Pareto's α and Gini's δ for incomes purport to be descriptive indices. An index, which has been descriptive for one characteristic, may be only global for another ; the α and δ indices have been descriptive for incomes, but only global for fortunes.

Descriptive Statistics. As statistics has two distinct meanings : (1) the plural of statistic and (2) methods of collecting, presenting and analysis data, so descriptive statistics can mean : (*a*) statistical data of a descriptive type, or (*b*) simply calculated measures of data, which are of a descriptive rather than an analytical type. The calculated statistics generally relate to the data on which they are based, rather than to a population from which the data may have been drawn.

Most writers on statistics confine the term to : (*i*) methods of presenting statistics, using tables, graphs, charts, pictograms etc. ; (*ii*) some simple methods of calculating measures of central tendency (*e.g.*, mean, median and mode) and measures of dispersion (*e.g*., standard deviation).

Deseasonalization. A statistical process used to remove the effects of seasonality from a time series.

De-Seasonalized Series. A time series the observations of which have been adjusted by removing the variation due to seasonal factors (the seasonal variation).

Design of Statistical Experiment. A statement of the purpose of, and proposed approach to an experiment or investigation involving statistical analysis. The approach covers the statistical methods to be used and a realistic form of mathematical model, which takes account of the relevant known and ascertainable basic and derived statistics.

Destructive Test. Under certain conditions it has been possible that the carrying out of an inspection test on a manufactured product will cause destruction of the particular test specimen. This is termed as a destructive test. For example, a test of the fusing current in an electric has been of such a kind and many of the inspection tests for explosive products must necessarily be destructive. Under such conditions there has been every incentive to design sampling schemes which minimise the number of items to get tested.

Determination, Coefficient of. Refers to the square of the product-moment correlation between two variates, r^2, so-called because it represents the proportion of the variance of one variate, y, given by the other, x, when y is expressed as a linear regression on x.

More generally, if a dependent variate is having multiple correlation R with a set of independent variates R^2 is termed as the coefficient of determination.

The quantity is also termed as the index of determination.

Deterministic Distribution. Refers to a degenerate distribution which consists of a mass concentrated at a single point $x=k$ so that its probability function has been

$$\Pr(x)=\begin{cases}1 & x=k\\ 0 & x\neq k\end{cases}$$

Deterministic Model. A deterministic as model, as opposed to a stochastic model, may be defined one which contains no random elements and for which, therefore, the feature course of the system has been determined by its position, velocities, etc., at some fixed point of time.

Deterministic Process. Refers to a stochastic process having a zero error of prediction ; one in which the past completely ascertains the future of the system.

Deviance. A term which was proposed by M.G. Kendall to and used to denote the sum of squares of observations about their mean.

Deviate. Refers to the value of a variate which is measured from some standard point of location, usually the mean. It is generally understood that the value could be expressed in standard measure, *i.e.*, as a proportion of the parent standard deviation.

Deviation. Refers to the difference between a value of a variable and the mean of its distribution.

Example : The numbers 3, 6, 7, 10 and 4 are having mean 6, and so their deviations are :

$$3-6=-3$$
$$6-6=\ 0$$
$$7-6=\ 1$$
$$10-6=\ 4$$
$$4-6=-2$$

The sum of all the deviations in such a set should be zero.

Deviation has been used in the calculation of variance, standard deviation, covariance, etc., but has been not often used in its own right.

Diagonal Regression. If two variables are subject to errors of observation and get connected by a linear relation, a set of observed values will not, in general, lie on a straight line. The linear relation is estimated by regression techniques (although, strictly speaking, the problem is quite different from the determination of regression lines) ; and this method is having the embarrassing property of yielding two different regressions. Frisch (1934) gave compromises between these two in the form of 'Orthogonal Regression' and diagonal regression. The latter has been a line which is lying between the regression lines and passing through their point of intersection, and having the equation

$$\frac{Y-m_y}{\sigma_y}=\frac{X-m_x}{\sigma_x}$$

where m_y, m_x denote the means add σ_y^2, σ_x^2 the variances of y and x respectively. If $\sigma_y=\sigma_x$ the diagonal regression coincides with the orthogonal regression. Neither, strictly speaking, has been a regression.

Diagrammatic Presentation. Presentation of statistics, by chart on a wall for general information, or in a book or artical to illustrate an argument.

Dichotomous Population. In statistics, a population that has been divided into two mutually exclusive classes. Thus all cases in one category have a certain characteristic or more than a certain magnitude of a characteristic and no cases in the other category have that characteristic or that magnitude, for example, a population divided into male and female, or over 21 years of age and under 21.

Difference and sum of two or More Random Variables. Refers to those variables which have been the difference and the sum of two or more random variables. If two random variables, X_1 and X_2 are added or subtracted to produce a third random variable, its mean and variance would be given by these relationships.

Mean $(X_1+X_2)=$ Mean $(X_1)+$ Mean (X_2) or $E(X_1+X_2)=E(X_1)+E(X_2)$

Variance Var $(X_1+X_2)=$ Var $(X_1)+$ Var $(X_2)+2$ cov (X_1+X_2)

Mean $(X_1-X_1)=$ Mean $(X_1)-$ Mean (X_2) or $E(X_1-X_2)=E(X_1)-E(X_2)$

Variance Var $(X_1-X_2)=$ Var $(X_1)+$ Var $(X_2)-2$cov (X_1, X_2)

where E denotes expectation, cov denotes covariance and Var denotes variance. If X_1 and X_2 have been independent variables then cov $(X_1, X_2)=0$. In that case

Var $(X_1+X_2)=$ Var $(X_1)+$ Var (X_2)

Var $(X_1-X_2)=$ Var $(X_1)+$ Var (X_2)

Difference Sign Test. When observations are considered on a phenomenon which a moving through time and generates an ordered set, *i.e.*, a time series, one useful test against trend, etc., would be the difference sigh test. This consists of counting the number of positive first differences of the series, that is to say, the number of points where the series increases. It has been used especially as a test against linear trend and as such has been superior to the Turning Point test but inferior to tests based on rank order.

Diffusion Index. A term which was proposed by Burns (1950) and Moore (1950) to deenote the proportion of a set of time series in a given collection of series which have been increasing at a given point of time.

Diffusion Process. A type of Additive Process which describes certain kinds of diffusion. The process has been such that the 'displacement' of the variate (its increment) in time *dt* follows a Normal distribution with variance proportional to *dt*.

Dimroth-Watson's Distribution. It is a spherical distribution of the symmetric girdle type which was introduced independently by Dimroth (1962, 1963) and Watson (1965). Its probability density function would be as follows :

$$f(l, m, n)=\{b(K)/2\pi\} \exp \{-K(l\lambda+m\mu+n\nu)^2\}'$$
$$-\infty<K<\infty$$

where (l, m, n) and (λ, μ, γ) denote the vectors of direction cosines of a three-dimensional directional random variable and its mean direction respectively, and

$$b(K)=1\Big/\left(2\int_0^2 e^{-kt^2}\,dt\right).$$

Direct Probability. Refers to an expression which is assumed to be antithetical to inverse probability, although neither expression has been very logical. It usually represents probability when used to proceed from the given probabilities of prior events to the probabilities of contingent events ; *e.g.*, if it has been given that the probability of throwing each number with an ordinary six faced die has been $\frac{1}{6}$, the probability of throwing a score of 15 in three throws is directly ascertainable. The relation has been analogous to the deductive relations of logic.

Direct Sampling. This term is used when the sample units have been the actual members of the population and not, for example, some kind of record relating to such numbers, like census form, ticket or registration card. The term has been related to the directness of the observation of the units which enter into the sample, not to the process by which they have been selected.

Dirichlet Distribution. Refers to the *k*-variate analogue of the beta distribution.

Disarray, Coefficient of. Two rankings A and B may get transformed one into the other by a successive interchange of pairs of neighbouring items in one of term. There would be a minimum number of moves in which the operation could be made and a coefficient of disarray could be constructed from it. This coefficient has been in fact the same as Kendall's tau.

Discontinuous Variate. Refers to a variate which may take only a discontinuous set of values.

Discrepance. In variance-analysis, this term has been used by some writers to denote the error sum of squares.

Discrete. Consisting of distinct or separate parts.

Discrete Circular Uniform Distribution. See circular lattice distribution.

Discrete Data. Data that do not progress from one class to the next without a break ; *i.e.*, where classes represent distinct categories or counts and may be represented by whole numbers.

Discrete Lognormal Distribution. It is a confusing term which is used to denote a distribution obtained as the compound of a Poisson distribution with the parameter λ distributed lognormally.

Discrete Normal Distribution. It is a discrete analogue of the continuous Normal distribution of the form given below :

$$\Pr(x) = A \exp\left\{-\frac{(x-m)^2}{2v}\right\}$$

where $x=0, \pm 1, \pm 2, \ldots$, and $m \approx \mu$ with $v \approx \sigma^2$ and

$$1/A = \Sigma \exp\left\{-\frac{(x-m)^2}{2v}\right\}$$

with v and $m^2/v \geqslant 10$.

Discrete Pareto Distribution. It is the discrete analogue of the continuous Pareto law :

$$\Pr(x) = \frac{1}{\zeta(s)} \frac{1}{x^s} \quad x=1, 2, \ldots, \infty$$

where ζ denotes the Riemann zeta function.

Discrete Power Series Distribution. It is a general probability distribution (Noack, 1950) of the form given below :

$$\Pr(x) = a_x \theta^x / f(\theta) \text{ with } x=0, 1, 2, \ldots \text{and}$$

$$f(\theta) = \sum_{x=0}^{\infty} a_x \theta^x.$$

By appropriate choice of $f(\theta)$ it is possible to derive several common discrete distributions ; for example, the Poisson, negative binomial, Polya-Eggenberger.

Discrete Probability Distribution. A probability distribution in which the variable is allowed to take on only a limited number of values.

Discrete Probability Law. A probability law is termed as discrete if it corresponds to a discrete distribution function.

Discrete Random Variable. A random variable that is allowed to take on only a limited number of values.

Discrete Rectangular Distribution. Refers to a distribution (of finite range) in which each of a discrete set of x variables, n in number, is having the same probability $1/n$. It has been mostly used for equi-spaced variate values, *e g.*, the numbers 0—9 if equally probable, from a discrete rectangular distribution.

Discrete Series. This term is used in the literature in two senses : (1) a series which is naturally discrete, *i.e.*, one whose values fall naturally at a number of fixed exact measurements, *e.g.*, a number of fixed integers ; and (2) a series which is expressed in discrete form by using class measurements, *i.e.*, class intervals.

Discriminatory Analysis. For a set of multivariate observations on samples, known with certainly to come from two or more populations, the problem has been to set up some rule which would allocate further individuals to the correct population of origin with minimal probability of misclassification. This problem and sundry elaborations of it would give rise to discriminatory analysis.

Dishonest Process. It is an alternative name for a logical process, and the opposite in a mathematical sense, of an 'honest' process.

Disjoint, ***abj.*** **(of Sets) Non-Overlapping.** In the Venn diagram (Fig.) the sets A and B are disjoint, $A \cap B = \phi$.

The equivalent term for events (*i.e.*, describing two events that cannot both occur) has been exclusive.

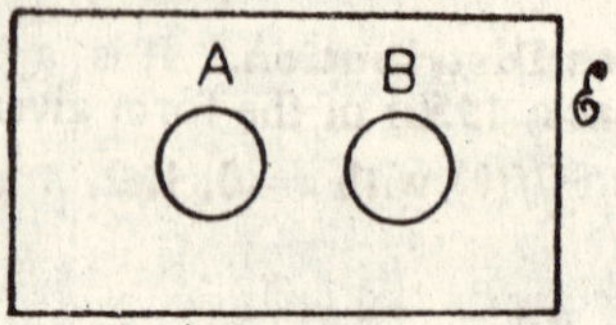

Disjoint. A Venn diagram of disjoint sets A and B.

Dispersion. Refers to the spread of a distribution. In the diagram distribution B is having greater dispersion, or spread than A although they both are having the same mean.

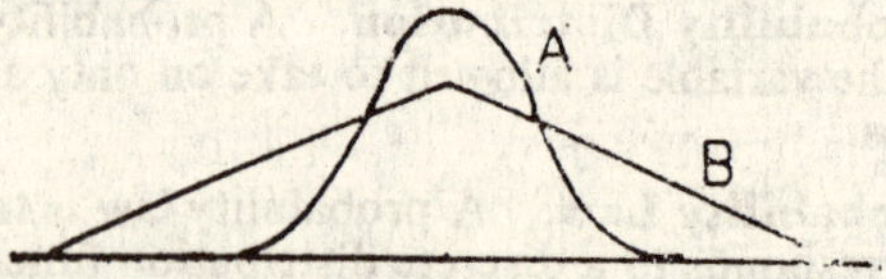

Dispersion. Diagram of two distribution with the same mean, but different dispersions.

There have been several measures of dispersion. Mean absolute deviation, standard deviation and semi-interquartile range have been all measures of how far above or below the middle a typical member might be found. Standard deviation has been the most important statistically and also, for small data sets, the most tedious to calculate by hand ; it gives more weight to extreme values than do the other two.

Disproportionate Allocation. Occurs in stratified sampling whenever different sampling fractions are used in selecting the sample in different strata. The most commonly used form of disproportionate allocation is optimal allocation.

Disproportionate Stratified Sampling. Stratification denotes the division of the population into subdivisions, known as strata, in each of which sampling is carried out independently. Once the strata have been formed, the investigator has complete freedom to choose the sampling fraction (or proportion of the elements to be selected) in each stratum. Disproportionate allocation (using different sampling fractions in different strata) will be justified on three counts : (1) the more dispersed or variable the elements are in a stratum the higher the sampling fraction ; (2) the more expensive it is to obtain information in a stratum the lower the sampling fraction ; (3) the more precision we require for the estimates in a stratum the higher the sampling fraction. Disproportionate allocation does, however, involve some risks. Whereas proportionate allocation cannot in general lead to a fall in precision compared to simple random sampling, disproportionate allocation—if based on faulty information about costs or dispertion—can lead to serious losses in precision. The additional costs in the analysis caused by the additional complexity of the sample design should be considered when deciding whether to use proportionate or disproportionate allocation.

Distribution. Refers to the set of frequencies (***observed distribution***) or probabilities (***theoretical distribution***) which are assigned to a set of events.

Example : ***frequency distribution*** (in which frequencies are given). The length of the reigns of English kings and queens, 827-1952 ad.

Length (Years)	<5	$5\leqslant<10$	$10\leqslant<15$	$15\leqslant<20$	$20\leqslant<25$
Frequency	11	11	8	8	7
Length (Years)	$25\leqslant<30$	$30\leqslant<35$	$35\leqslant<40$	$40\leqslant<45$	
Frequeney	4	1	6	1	
Length (Years)	$45\leqslant<50$	$50\leqslant<55$	$55\leqslant<60$	$60\leqslant<65$	
Frequency	0	1	2	1	

Example : Probability distribution (in which probabilities are given). The probability distribution of different totals when two dice are thrown. The top line (below) denotes the possible combined scores of two dice. The bottom line gives the probability of each combined score occurring.

Total score	2	3	4	5	6	7	8	9	10	11	12
Probability	$\frac{1}{36}$	$\frac{2}{36}$	$\frac{3}{36}$	$\frac{4}{36}$	$\frac{5}{36}$	$\frac{6}{36}$	$\frac{5}{36}$	$\frac{4}{36}$	$\frac{3}{36}$	$\frac{2}{36}$	$\frac{1}{36}$

Figs. 1, 2, 3, 4, 5, *a* and *b* are showing the graphs of some common distribution.

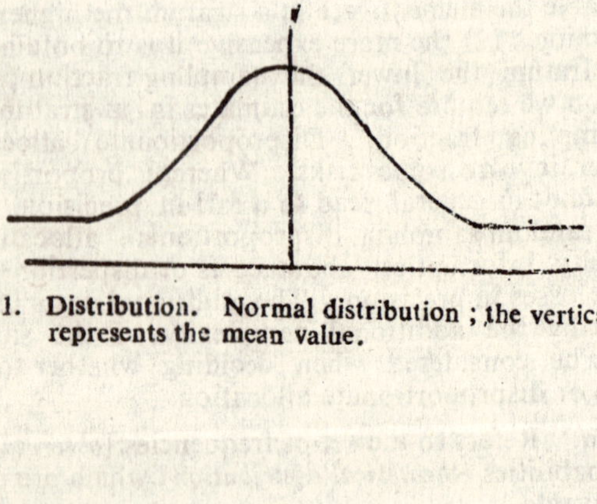

Fig. 1. Distribution. Normal distribution ; the vertical line represents the mean value.

Fig. 2. Distribution. Poisson distribution ; in this example, the mean μ is 2, so the distribution is denoted P (2).

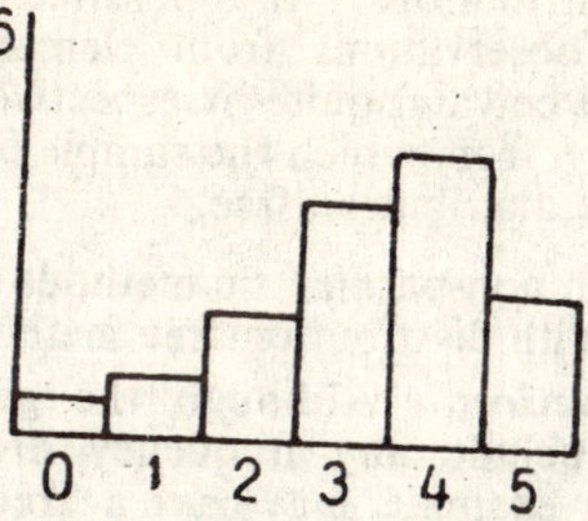

Fig. 3. Distribution. Binomial distribution ; in this example, the number of trials *n* is 5, the probability of success *p* is 7. Thus the distribution is denoted B (5,. 7).

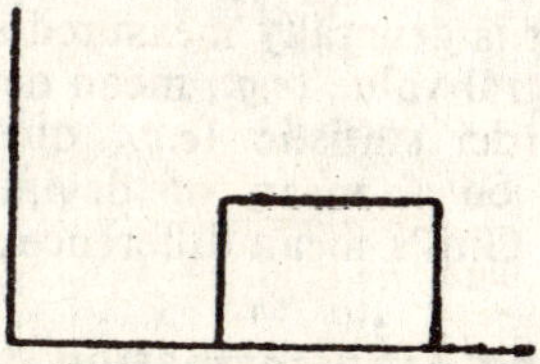

Fig. 4. Distribution. Uniform distribution

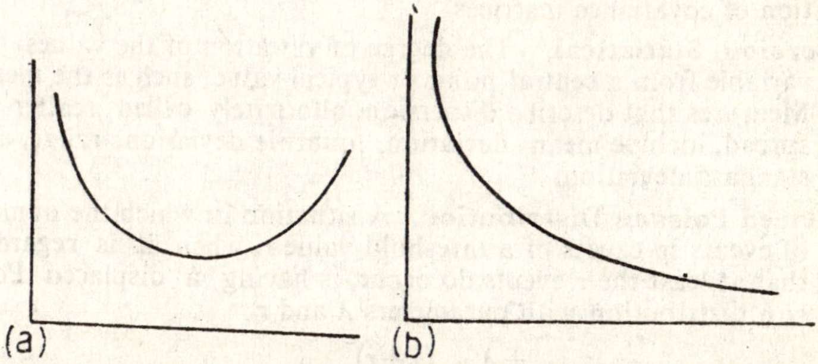

Fig. 5. Distribution. Empirical distributions described by their shapes. (*a*) U-shaped distribution. (*b*) J-shaped distribution.

Distribution Curve. The graph of a distribution function, showing frequency values plotted against an ordinate and variate values against an abscissa. The term is popularly used to denote a cumulative frequency curve.

Distribution-Free Methods. Methods of testing given hypothesis, which do not depend on assuming a particular parametric frequency distribution, but which are applicable to all frequency distributions. It is possible to conduct tests involving ranked observations from elements of a sample, whose results would be valid quite irrespective of the parameters of the population from which the sample is drawn. Such methods of testing are distribution-free.

The term non-parametric methods is sometimes used interchangeably with distribution-free methods.

Distribution Function. Although the general meaning of this term may include any frequency distribution, where a given variate x, for example, may have a frequency 5 at x_1, 10 at x_1, 12 at x_3, etc., the term is often specifically used to mean the total (or cumulative) frequency of components of a series with variate values less than or equal to x_i.

Dispersion. Refers to the degree of scatter which is shown by observations. It is generally measured as an average deviation about some central value (*e.g.*, mean deviation, standard deviation) or by an order statistic (*e.g.*, quartile deviation, range) but could also be a mean of deviations of values among themselves (*e.g.*, Gini's mean difference and also standard deviation).

Dispersion Stabilising Transformation. A term introduced by Ruben (1966) to denote a generalisation of the variance stabilishing transformation. The generalisation covers the stabilisation of covariance matrices.

Dispersion, Statistical. The degree of variation of the values of a variable from a central point or typical value, such as the mean. Measures that describe dispersion, alternately called scatter or spread, include mean deviation, quartile deviation, range, and standard deviation.

Displaced Poisson Distribution. A situation in which the number of events in excees of a threshold value r, when it is regarded that at least the r events do occur, is having a displaced Poisson distribution with parameters λ and r

$$Pr\ (x) = \frac{e^{-\lambda}\ \lambda^{(x+r)}}{I\ (r,\lambda)\ (x+r)!}$$

Distance. This word is finding use in many statistical contexts in its ordinary sense, *e.g.*, the 'distance' of a value x from some origin a is $x-a$. A specialised use is found in the notion of 'distance' between two variates, x and y, which is defined as the expected value of $x-y$; or the 'distance' between two populations, which is defined as the difference of their means (though other definitions are possible). [See also **Bhattacharyya's distance, D² statistic**].

Distance Distributions. Refer to distributions which arise from more or less naturally defined geometrical systems of variation. For example, if two points have been taken at random within a circle, with unit radius and centre at the origin, the joint distribution of the coordinates (X_1, Y_1) (X_2, Y_2) would be as follows :

$$Pr\ x_1, y_2\ ;\ x_2, y_2\ ^{(x1,\ y1\ ;\ x2,\ y2)}$$
$$=\pi^{-2}(0 \leqslant x_j^2 + y_j^2 \leqslant 1\ ;\ j=1,2)$$

Distance Measure. A measure of dispersion in terms of the difference between two values in the data set.

Distribution Curve. It refers to the graph of cumulated frequency as ordinate against the variate value as abscissa, namely the graph of tne distribution function.

Distribution-Free Method. It is a method, *e.g.*, of testing a hypothesis or of setting up a confidence interval, which does not depend on the form of the underlying distribution ; for example, it is possible to obtain confidence intervals for the median, based on binomial variation, which are valid for any continuous distribution. In the case of testing hypotheses, it means that the test is independent of the distribution on the null hypothesis. Distribution-free inference or distribution-free tests are sometimes termed as non-parametric but this usage has been confusing and should be avoided. It has been better to confine the word 'non-parametric' to the description of hypotheses which do not explicitly make an assertion about a parameter.

Distribution Function. Refers to the function $F(x)$ which gives the cumulative frequency or the cumulative probability of the random variable X. Thus, $F(x)$=Frequency (or probability) of $X \leqslant x$ for a set of values of X.

Distribution of Run Length. Refers to the distribution of lengths of runs of attributes in a series. A run may be defined as a sequence of items of the same kind which get terminated by one or more items of another kind.

Disturbed Harmonic Process. It is a form of stochastic process which was proposed by Yule to explain the continual change of amplitude and shift of phase which appears to be typical of time series from the economic and meteorological sciences. If a series has been observed at points $t=1, 2, \ldots$ an ordinary harmonic movement could be expressed by the equation

$$u_{t+1} - 2u_{t+1} + u_t = 0.$$

If the zero is replaced by a random term ϵ_{t+2} on the righthand side the motion is said to be that of a disturbed harmonic. It has been a limiting case of an auto-regressive process but has been not stationary.

Dixon's Statistics. Refers to a class of statistics for the rejection of the largest observation proposed by Dixon (1950, 1951) of the form

$$r_{ij}=\frac{x_{(n)}-x_{(n-i)}}{x_{(n)}-x_{(j+1)}} \qquad \begin{array}{l} i=1, 2 \\ j=0, 1, 2. \end{array}$$

where $x_{(k)}$ implies the kth value of a sample on n when ranked by size.

Domain of Study. A domain is a part of the population in which the investigator has a special interest and for which separate estimates are desired. Stratification, where possible, may be used to separate the domains of study so that the desired precision can be built into the survey design. Although the term should be used to denote subdivisions of the population for which estimates with known precision are planned, it is sometimes used to specify subdivisions for which separate estimates are to be calculated but for which the requisite prior information for the design in not available. The term subclass would be more appropriate in these cases.

Dominating Strategy. Refers to a strategy which has been in no respect worse than all others and in some respect has been better.

Double Binomial Distribution. This distribution has been a weighted average of two binomials having parameters p_1 and p_2 ; $p_1 < p_2$ and weights w_1 and w_2 ; $w_1 + w_2 = 1$.

Double Confounding. This term means confounding of two different groups of treatment contrasts having two different sources of variation in an experimental design. Thus in the row and column layout of a quasi-latin square certain interactions get confounded with rows, and certain others with columns.

Double Exponential Distribution. It is the distribution having frequency function $f(x)=ae^{b|x-c|}$, $-\infty \leqslant x \leqslant \infty$, a, b and c being constants, $b<0$. The distribution may be considered to be an ordinary exponential together with its reflexion about the point c. An example of the double exponential distribution has been that of the distribution of the mid-range from a rectangular population.

Double Exponential Regression. It is an expression which is sometimes used for a regression equation in which the dependent variable (y) has been a lidear combination of two simple exponential terms in (x).

Double Hypergeometric Distribution. If, from a finite population having two kinds of item, a random sample has been drawn without replacement and then a second random sample without replacing the first, the distribution of the (x_1, x_2) successes would take the form of the double hypergeometric.

Double Poisson Distribution. It is a distribution in which the parameter λ of the Poisson series is itself considered as distributed in the Poisson form. The distribution arises particularly in ecology where the numbers offspring of parents which have been themselves distributed over space in the Poisson form have been also distributed in that form in the neighbourhood of the parent.

Double-ratio Estimator. Refers to an estimator which is built up from four variates, x_1, x_2, y_1, y_2 by using the ratio of the ratios y_1/x_1 and y_2/x_2.

Double Reversal Design. Refers to an extension of the switch-back or reversal design to the case of two treatments over four periods : A_1 B_2 A_3 B_4 for one-half of the experimental units and B_1 A_2 B_3 A_4 for the other half ; the suffixes have been the time periods. This design permits trends in the responses which are independent of the treatment effects.

Double Sampling. It refers to a standard form of sample design for industrial inspection purposes. According to the characteristics of a particular plan, two samples have been drawn, n_1 and n_2, and the first sample inspected. The batch can then be accepted or rejected upon the results of this inspection or the second sample be inspected and the decision made upon the combined result.

Doubled-tailed Test. It is a test for which the region of rejection includes areas at both extremes of the sampling distribution of the test function. It is general but not essential to allot one-half of the probability of rejection to each extreme, yielding a symmetrical test.

Doubly Noncentral F-distribution. Refers to the distribution of the ratio of two noncentral X^2.

Doubly Stochastic Poisson Process. It is a point stochastic process where the rate of occurrence in a Poisson process has been itself a stochastic process.

Doubly Truncated Normal Distribution. It is a distribution of Normal (Gaussian) form but with the variable not extending to infinity, being truncated below and above at specified points.

Downward Bias. It refers to a bias which tends to reduce a magnitude below its true value. In index number theory the expression is frequently found but has a rather obscure meaning owing to the somewhat arbitrary nature of the definition of bias in index numbers.

Dual Process. For every queueing process it has been possible to create an associated process by interchanging the distributions of inter-arrival and service times respectively : this associated process is known as the dual process by Takacs (1965).

Dual Theorem (Dual Problem). A term used in linear programming to indicate a reverse problem to its counterpart, often styled the primal problem.

Dummy Observation. In variance-analysis with disproportionate sub-class numbers it has been sometimes possible to get good approximations to the correct analysis by adding 'dummy' observations. These observations are usually inserted with values equal to the means of the particular cells concerned.

Dummy Variable. It is a quantity which is written in a mathematical expression in the form of a variable although it represents a constant ; for example, in a regression equation

$$y=\beta_0+\beta_1X_1+\beta_2X_2+\cdots+\beta_pX_p$$

it has been more convenient to attach to the coefficient β_0 a dummy X_0 which has been always unity so that the expression may be written

$$y=\sum_{i=0}^{p}\beta_iX_i.$$

Duo-Trio Test. It refers to a test in which three objects, two of which are alike are presented to a judge who tries to select the dissimilar object after being given the identity of two of them.

Duplicate Sample. A sample which is collected concurrently, *i.e.*, in the course of the same sample survey under comparable conditions, with a first sample. It acts in much the same way as a replication except that in some cases the only thing which can get replicated has been the act of taking a second independent sample.

Duplicated Sample. It is a sample which has been taken up twice for enquiry by two different parties of investigators. Two interpenetrating samples assigned to two investigating parties sometimes get so arranged, for purposes of control of field operations, as to have some common sample units these common units would constitute a duplicated (sub.) sample. This should get distinguished from duplicate sample.

Durbin-Watson Statistic. One of the assumptions of the ordinary least-squares method of estimating regression coefficients is that the disturbance (stochastic) terms are uncorrelated with one another. If the data used are recorded sequentially through time (time series data) then it often happens that these terms in the regression model are autocorrelated. A test for autocorrelation has been devised by Durbin and Watson. If $e_1, e_2, \ldots e_n$ are the unexplained residuals from the fitted regression equation the Durbin-Watson statistic is

$$d = \frac{\sum_{t=2}^{n} (e_t - e_{t-1})^2}{\sum_{t=1}^{n} e_t^2}$$

and its value lies between 0 and 4. If the model stochastic terms are uncorrelated (at lag 1) d will take values near 2; smallar values indicate positive autocorrelation and higher values negative.

Dvoretsky's Stochastic Approximation Theorem. It is a general theorem on the convergence of transformations with superimposed random errors (Dvoretsky, 1956). It is applicable to the stochastic approximation procedures of Robbins-Munro and Kiefer-Wolfowitz.

Dynamic Model. In econometrics a model is considered to be dynamic if it is having either or both of these properties : (1) at least one variable occurs in the structural equations with values taken at different points of time or in the form of time-derivatives, etc. ; (2) at least one equation contains a function of time.

Dynamic Programming. It is a method which is used to secure a sequence of consistently optimum solutions to multistage decision problems, sequential decision considering in fact.

Dynamic Stochastic Process. It is a class of second order process which was introduced by stigun (1903) and combine a wide sense stationary process and a determinition-station any process in a single representation.

E

Econometrics. Definitions of broad areas of knowledge and study have been necessarily somewhat loose as one subject area frequently overlaps into several others. This has been certainly true of econometrics, which, broadly speaking, is concerned with quantifying economic theories and relationships. Thus it plays a role to economics that is in many ways similar to that of the engineer in relation to the pure scientist.

Econometric Model. Refers to a set of one or more structural equations whose form (*i.e.*, which variables depend on which) gets derived from economic theory and in which any unknown parameters are estimated statistically. There have been normally as many structural equations in the model as there are endogenous variables that the model seeks to explain. Each equation relates one endogenous variable to other endogenous and exogenous variables. A simple example has been the two-equation model of the market for an industrial product consisting of supply and demand questions. The endogenous variables in this case would be price and quantity, and exogenous factors might include advertising expenditure (affecting demand) and research and development (*r & d*) expenditure (affecting supply). Thus we might have

Quantity $= a + b \times$ price $+ c \times (r \,\&\, d) + u_1 \rightarrow$ supply equation

Quantity $= d + e \times$ price $+ f \times$ advertising $+ u_2 \rightarrow$ demand equation

These may be written as :

$$y = a + bx_1 + cx_2 + u_1$$
$$y = d + ex_1 + fx_3 + u_2$$

where x_1 denotes price, x_2 denotes research and development, x_3 represents advertising and where a, b, c, d, e, f and coefficients to be estimated and u_1, u_2 are unexplained residual terms. Given values for the exogenous variables which can be fixed by the firms, we can solve the equations simultaneously to find values for price and quantity to be sold.

Edgeworths' Series. It is a method for representing certain skew frequency distributions and is based upon an expansion in terms of the Normal distribution and its derivatives. It is very similar to the Gram-Charher series.

If the cumulants of the distribution have been represented by k_1, k_2, etc., the Normal function $\frac{1}{\sqrt{(2\pi)^e}} \frac{1}{2^x}{}^2$ by $a(x)$ and the process of differentiating $\frac{d}{d_x}$ by D, the Edgeworth expansion of a frequency function $f(x)$ with zero mean and unit variance could be but as follows.

$$f(x)=\exp\left\{\sum_{j=3}^{\infty} \frac{(-D)^j}{j!} k_j\right\} a(x)$$

where the exponential has been expanded as a power series before the operations with D have been carried out.

Effective Range. A group of observations may have a limited number of outlying observations at either or both ends of the range. It is possible to obtain a very rough measure of dispersion by taking the 'effective' range, *i.e.* the range after these outlying values have been removed. The removal may have to be a matter of subjective judgment and inferences based or effective range have been somewhat doubtful value ; in fact the term itself has been not a good one.

Efficiency Equivalence. It refers to a broad extension of the concept of asymptotic relative efficiency to the case of regression parameters in second order processes.

Efficiency Factor. The efficiency factor of an experimental design may be expressed in terms of a ratio of error variances, or, from a slightly different viewpoint, in terms of the size or number of replications required to attain a given precision.

Efficiency of an Estimator. For an estimator which is asymptotically (for large samples) normally distributed and consistent, efficiency is defined as the ratio of sample sizes required to make its sampling variance equal to that of the efficient estimator as the sample size increases. Since sampling variances will depend on sample sizes this is just a way of assessing just how good a particular estimator is relative to the best one available for large samples.

Efficiency, Statistical. The extent to which a statistic, or value obtained for a given sample, provides the maximum possible amount of information about one or more characteristics of a population.

Efficient Estimators. Most common estimators have asymptotically (for large sample sizes) normal sampling distributions. Consistent estimators will also be asymptotically unbiased. Among such estimators the one with the smallest variance in large sample is called an efficient estimator. This acts as a standard against which the usefulness of other estimators may be judged.

Eisenhart Models. In variance analysis it has been customary to draw a distinction between type 1, in which the classification variables have been fixed, and type 2 in which they have been themselves random variables. The distinction has been pointed out by Eisenhart whose name has been frequently attached to the models, especially in balanced incomplete block experiments.

Element. A member of a group, set or collection of items which cannot be further subdivided into constituents that could be regarded as members of a collection. Hence the mathematical and statistical usage of the word is analogous to its usage in the physical sciences.

Elementary Unit. Refers to one of the individuals which, in the aggregate, compose a population : the smallest unit providing information which, by suitable aggregation, gives rise to the population property under investigation.

Element Sampling. Refers to a sampling scheme in which the sampling unit contains only one element. This is in contrast to cluster sampling, in which groups of elements are selected together.

Ellipticity. A term which is sometimes used to describe a bivariate normal distribution in which the variances have been not equal, leading to the idea of a 'test of ellipticity', *i.e.*, a test of the inequality of the variances.

Elliptically Symmetric Distributions. This name is assigned to distributions whose density functions depend only on quadratic functions of the variables so that contours of constant density are elliptical. They are including spherically symmetric distributions as a special case.

Empirical. Derived from or relating to experiment and observation, rather than theory. An example of an empirical law has been the Rank Size Rule, which is used in geography. This states that if P_1 is the population of a country's largest city (or town), then P_1/n is that of its nth city. There has been no theoretical basis for this rule ; it has just been found to work in many cases, as in the case of the USSR (1979 figures).

Rank	*City*	*Actual size (× 1000)*	*Predicted size (× 1000) (Rank Size Rule)*
1	Moscow	8,011	8,011
2	Leningrad	4,588	4,006
3	Kiev	2,144	2,670
4	Tashkent	1,779	2,003
5	Baku	1,550	1,602
6	Kharkov	1,444	1,335

Source : *Philip's Certificate Atlas*, 1978.

Empirical Distribution Function. Given an ordered sample of n independent observations.

$$x_{(1)} \leqslant x_{(2)} \leqslant x_{(3)} \leqslant \ldots \geqslant x_{(n)},$$

the function $S_n(x)$ defined as :

$$S_n(x) = \begin{cases} 0, & x \leqslant x_{(1)} \\ \dfrac{k}{n}, & x_{(1)} < x < x_{(k+1)} \\ 1, & x_{(n)} < x \end{cases}$$

is termed as an empirical distribution function.

Empirical Probit. The analysis of experimental results concerned with the relationship between levels of a stimulus and the responses thereto has been generally carried out in terms of the percentage of test subjects reacting to the stimulus. The analysis often gets facilitated by transforming the percentages into probits. The probits corresponding to these observed percentage are termed as empirical probits, in contradistinction to expected probits and working probits, which get related to a fitted regression line.

Employment Function. An equation or mathematical model expressing the average labour force employed as a function of the scale of production, past experience of the product or products being produced, and the average size or weight of the product. In effect it consists of a form of the production function, which expresses output in terms of quantity as a function of the average labour force, past experience and the average size of the product.

Empty Cell Test. It is a rather inefficient class of distribution-free tests which is characterised by the test statistic being the number of cells in a uni-or multivariate classification into which none of the sample members fall.

End Corrections. An expression which is used in several statistical contexts, to denote corrections made to extreme values. For example, in taking a systematic sample from material which varies continuously it may be possible to got some increase in accuracy of estimation by using corrections to the term at each end of the sample. Again, in fitting a curve to a time series, corrections could be made to end values to improve the fit. The expression is also found in the calculation of serial correlations where certain alterations become necessary to formulae relating to infinite series to allow for the fact that the observations are having finite length.

Endogenous. In multivariate analysis the variable can be divided into two classes endogenous and exogenous. Those endogenous to the system are those which are influenced by other variables within the system.

Endogenous Variable. This is a variable in an econometric model that is explained by the model. Thus it plays the same role as the dependent variable does in a regression equation. The model itself is a simplified representation of a complicated system in which many endogenous variables are being simultaneously determined. The driving force of the system is provided by the exogenous variables—those whose values are fixed without reference to the system under discussion.

Endogenous Variate. The statistical representation and analysis of multivariate systems all generally involving a primary division of variates into those which are endogenous and those which are exogenous. Endogenous variates have been those which form an inherent part of the system, as for instance price and demand in an economic model.

Ensemble. Refers to the infinite set of realisations of a single stochastic process. The concept has been frequently extended to apply to a sample set of realisations.

Ensemble Average. Given a number of independent sample realisations from the same generating process, it is possible to estimate the mean of the process by the overall mean of the samples. The estimator is termed as an 'ensemble average'. If the process has been stationary, its mean can be estimated by a simple average over time instead of by the 'ensemble average.

Entry Plot. A plot through which a serial cluster gets entered or determined.

Enumeration Data. Refers to data which is consisting of numbers of individuals in various categories.

Example : a flower is having three varieties, red, white and pink, 400 flowers are collected and found to be :

Red	89
Pink	201
White	110

There have been thus three categories, red, pink and white ; the data consist of the number of flowers in each.

Envelope Power Function. A power function which is based on a critical region which itself is obtained as the envelope of a number of critical regions.

Epsem Sampling. The word 'epsem' denotes an Equal Probability of Selection Method. Epsem sampling therefore includes any sampling scheme in which every element in the population has an equal probability (chance) of selection. The big advantage of epsem is that it leads to self-weighting samples, in which there is no need to give different weights to different elements in order to obtain good estimates of the population parameters.

Equal Probability of Selection Method. It is any method of sampling in which the population elements are having an equal probability of selection. 'EPSEM' sampling could result from either equal probability selection throughout, or, from variable probabilities that compensate each other in multi-stage selection.

Equal Spacings Test. It is a non-parametric test of uniformity for circular distributions which is invariant under rotation.

Equal-Tails Test. Refers to a symmetrical double-tailed test.

Equally Correlated Distribution. A multivariate distribution has been equally correlated if the correlation between each pair of variates has been equal to ρ.

Equilibrium. In queueing theory, systems in 'statistical equilibrium' have been those in which the number of customers or items waiting in the queue oscillates in such a way that mean and distribution are constant over a long period.

Equilibrium Distribution. A physical, economic or social system which gets settled down to a stable statistical behaviour is having a stationary, or equilibrium, distribution which specifies the limiting proportions spent in the designated states of the system.

Equiprobable. (Of values) of equal probability. The possible scores if a die is thrown, 1, 2, 3. 4, 5, and 6, are equiprobable, each is having a probability of occurring of 1/6. The scores 0 to 36 on a fair French roulette wheel, have been also equiprobable, at 1/37 each.

Equitable Game. It is an alternative name for 'fair' game.

Equivalent Samples. Any two samples which are selected by the same sampling scheme and contain the same set of d ($\leqslant n$) population members, yield the same information about the parameters, irrespective of whether these d members appear with different frequencies in the two samples.

Ergodicity. Generally, this term denotes a property of certain systems which develop through time according to probabilistic laws. Under certain circumstances a system will tend in probability to a limiting from which has been independent of the initial position from which it started. It is the ergodic property.

Erratic Movements. In a time series these are irregular movements as opposed to trend or cyclical components. They include the ordinary random fluctuations that continually occur but exclude longer departures than can be traced to specific and isolated incidents such as wars and strikes ; these latter are referred to as episodic movements.

Error. In general, refers to mistake or error in the colloquial sense. There may, for example, be a gross error or avoidable mistake ; an error of reference, when data concerning one phenomenon get attributed to another ; copying errors ; an error of interpretation.

Error, Alpha. The rejection of a true null hypothesis. Since null hypotheses are accepted or rejected on the basis of statistical probability, the possibility exists that a true null hypothesis may be erroneously rejected.

Error Band. In estimation or prediction the estimated or predicted value gets bracketed by a range of values determined by standard errors, confidence intervals or similar methods within which the value may be regarded to lie with a certain probability. This is termed as the error band.

Error, Beta. The retaining of a false null hypothesis. Since null hypotheses are accepted or rejected on the basis of tests of statistical probability, the possibility exists that a false null hypothesis may be accepted. The degree of danger of accepting a false null hypothesis depends upon the level of significance used in testing the hypothesis : the lower the level of significance used, the greater · the probability of making a beta error. Beta error is sometimes referred to as type II error. See error, alpha, hypothesis, null ; probability ; significance, statistical.

Error, Chance. In statistics, one of a number of errors that arise from chance and tend to compensate for one another and cancel each other out Chance errors, sometimes referred to as random or accidental errors, are usually contrasted with constant error or bias.

Error Curve. The normal distribution curve.

Error in Equations. An equation in variables or variates may be inexact either because the equation has been not a complete representation of the situation, as in a demand-supply equation which does not include other factors such as income or employment, or because it gets disturbed by extraneous sources of variation as in an autoregression equation.

Error Mean-Square. It is equal to the residual or error sum of squares divided by the number of degrees of freedom on which the sum gets based. It gives an estimator of the residual or error variance.

Error of Estimation. In general, refers to the difference between an estimated value and the true value.

Error of First Kind. If, because of a statistical test, a statistical hypothesis gets rejected when it has to be accepted, *i.e.*, when it has been true, then an error has been committed. This class of error is termed an error of the first kind and is fundamental to the theory of testing statistical hypotheses associated with the names of Neyman and E.S. Pearson.

Error of Observation. An error which arises from imperfections in the method of observing a quantity, whether due to instrumental or to human factors.

Error Sum of Square. The sum of squares of residuals between observations and estimates using a particular multiple regression model. In analysis of variance it is assumed that observations result (1) from the effect of a number of regressors whose variances can be estimated and (2) from a random component, *i.e.*,

$$y = \sum_{i=1}^{p} \beta_i x_i + \epsilon$$

where a linear model is used.

Error Variance. Refers to the variance of an error component. Thus, if the generating model of a set of data is having certain systematic components together with a stochastic component, the variance of the latter is the error variance.

Esseen's Lemma. This result was due to Esseen, (1945). This formula gives an estimate of the closeness of two distribution functions F and G in terms of the closeness of the characteristic functions on an arbitrary interval $-T, T$. It states that if F and G have been distribution functions with respective characteristic functions f and g, and if G is having first derivative G', then

$$\underset{x}{\text{Sub}} \mid F(x)-G(x) \mid \leqslant \frac{1}{\pi}\int_{-T}^{T} \left| \frac{f(t)-g(t)}{\pi t} \right| dt + $$
$$+\frac{24}{\pi T}\underset{x}{\text{Sub}} - G'(x) \mid .$$

Estimate, 1. Refers to an approximate calculation. 2. Refers to an approximate idea of something (*e.g.*, size, cost). 3. To make an approximate calculation, or to form an approximate idea of something (*e.g.*, size, cost).

Estimating Equation. A mathematical formula that relates the unknown variable to the known variables in regression analysis.

Estimator. A random variable, formula or procedure for estimating the value of a parent population parameter from sample data. Its value depends on the particular sample involved. An estimator should ideally be consistent, unbiased and efficient. Estimated values of population parameters are distinguished from their true values by the convention of using the symbol $\hat{\sigma}$. Thus σ=true parent population standard deviation while $\hat{\sigma}$=estimated (from a sample) parent population standard deviation. See also efficient estimator, linear estimator.

Euler Diagram. See venn diagram.

Event. In probability this term is employed to mean any one of a set of probable outcomes or happenings, which in simple illustrations of probability are always mutually exclusive. For example, if a die is cast, the outcome '6' is an event, which excludes the outcomes '1', '2', '3', '4' and '5'.

Event Space. See sample space.

Evolutionary Process. Refers to any non-stationary process. The probability distributions associated with the process have been not independent of the time.

Exact Chi-Squared Test. See Fisher-Yates test.

Exact' Statistical Method. It is used in interval estimation to imply that the probability distributions involved have been completely known and that probability levels quoted have been exactly achieved.

Exceedances Tests. Refers to a class of distribution-free tests of identity of two populations which are characterised by having a test statistic that is the number of observations in a simple random sample from one population that exceed a particular order statistic in a simple random sample from the other population. See exceedances, distribution of.

Exclusive. (Of two events) such that both events cannot take place. Thus events A and B are said to be exclusive if B cannot take place when A does, and *vice versa.* It has been however, possible that neither A nor B takes place.

Example :

A : getting a 6 when throwing a die

B : getting a 4 on the same throw of the die

When three or more events have been such that no two of them can take place at the same time, they are considered to be *mutually exclusive.* Compare disjoint.

Exhaustive. **1.** (Of two events) such that at least on must occur. Thus events P and Q are said to be exhaustive if either P alone, or Q alone, or both P and Q together, ***must*** take place it is not possible for neither to take place. Exhausive events cover all possibilities.

Example :

P : An athlete does a track event

Q : An athlete does a field event

As all athletics events have been track, or field, or both (*e.g.*, the decathlon), P and Q are exhaustive events. 2. (of a greater number of events) such that they cover all possibilities between them, they are comprehenesive in their scope.

Exhaustive Classification. A classification or categorization of the variate values which includes all possible values. If, in addition each possible value can belong to one and only one category the classification is said to be exhausive and mutually exclusive.

Exhaustive Sampling. This term is occasionally encountered. This term indicates that, in order to achieve the sample size necessary for the required precision, the sampling has included all of the population under study. That is to say that a complete enumeration rather than a sample is taken.

Exogenous. In multivariate analysis the variables can be divided into two classes—endogenous and exogenous. Those exogenous to a system are those determined outside the system and not influenced by other variables in the system (or such determination is not of interest in the model).

Exogenous Variable. In an econometric model an exogenous variable is one that is determined outside the model and is not itself jointly dependent with the endogenous variables explained by the model.

Exogenous Variate. This term is the opposite to an endogenous variate.

Expectation. Refers to the expected or mean value of a random variable, or of a function of the variable. Symbol : E ( ). For a discrete variable, X, taking values $x_1, x_2, ..., x_n$ with probabilities $p(x_1), p(x_2), ..., p(x_n)$, the expectation of X, E(X) or $\bar{X}$, would be given as follows :

$$\sum_{i=1}^{i=n} x_i p(x_i)$$

and for a continuous variable

$$\int x f(x) dx$$

where $f(x)$ denotes the probability density function of X.

The 'expected' value is not necessarily the most frequently occurring value or even a possible value ; *e.g.*, if a variate can take each of the values 0 and 1 with probability $\frac{1}{2}$ and no other value is possible, the expected value is $\frac{1}{2}$.

Expected Frequencies. The frequencies we would expected to see in a contingency table or frequency distribution if the null hypothesis is true.

Expected Value/Expectation. The mean value of an estimator or function in repeated sampling is its expected value. Thus, if an estimator could take the values 3, 7, 12, 2, 15, 9 each with probability 1/6, the expected value of the estimator is 1/6[3+7+12+2+15+9]=48/6=8. Thus the expected value need not be a value which itself occurs at all. For example, with a fair die (each side having equal probability of occurring in a single toss) the expected value is $1/6[1+2+3+4+5+6] \times 3\frac{1}{2}$, which can never occur. The symbols E[ ], ϵ[ ] and EV[ ] are commonly used to denote expected value.

Expected Value Criterion. A criterion requiring the decision maker to calculate the expected value for each decision alternative (the sum of the weighted payoffs for that alternative in which the weights are the probability values assigned by the decision maker to the states of nature than can happen).

Expected Value of a Random Variable. The sum of the products of each value of the random variable with that value's probability of occurrence.

Experiment. Any process which results in the collection of data. (This meaning for the word has been somewhat different from that usually given to it in other scientific disciplines.)

Examples: Of experiments in a statistical sense are carrying out a survey; measuring the daily rainfall; recording the longevity of animals of a particular type.

Experimental Design. Refers to the way in which a statistical experiment is planned. The example which follows illustrates the use of a number of different designs, namely: *completely randomized design*, *randomized blocks design*, *Latin square*, and *Graeco-Latin square*.

Experimental Error. Refers to the difference between the value of a quantity found in an experiment and that which should have been found had the experiment been conducted perfectly. Experimental error has been due to experimental technique, and does not include mistakes in mathematics or experimental design.

More specifically, the term experimental error is used to denote the essential probabilistic variation to be expected under repetition of the experiment, not actual mistakes in design or avoidable imperfections in technique. It has been the aim of good experimental design to provide valid measures of the experimental error in the more restricted sense.

Experimental or Empirical Probability. Refers to a probability which is determined by experiment or observation rather than from some underlying theory. The figures used by life insurance companies in calculating the risks for particular individuals are included into this category. On the other hand, a skilled bridge player decides his strategy on the basis of theoretical probabilities.

Explosive Process. It is a stochastic process which is having no bound to the expectation of the mean square. A process the values of which may increase without limit (in absolute magnitude, so that oscillations are possible) as time goes on. The term is not to be recommended.

Exponent. The power to which a number is raised. Hence if a number is squared the exponent is 2. In the case $10^{8.8}$ the exponent is 8.8.

Exponential. (Of a value or function) raised to the power of. Thus the function $F(x)=a^x$ has been an exponential function ; x has been called the *exponent*.

The term exponential has been most commonly used when the exponent has been the power of e, the base of natural logarithms. In this case the exponential of x has been e^x, which has been sometimes written as $\exp(x)$.

Exponential Distribution. It is a continuous distribution with probability density function given by :

$$f(x)=\lambda e^{-\lambda x}$$

(see Fig. 1)

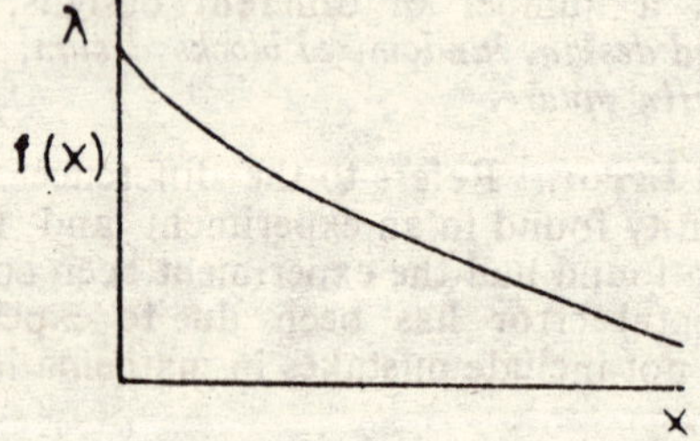

Fig. 1. Exponential distribution. For this distribution :

$$\text{Mean}=\frac{1}{\lambda} \quad \text{Variance}=\frac{1}{\lambda^2}$$

The exponential distribution has been sometimes used in connection with the length of life of certain materials.

Exponential Regression. A term which is used to denote a relationship of the form.

$$y_i=a+\beta e^{\lambda x t}+\epsilon$$

where ϵ denotes a random residual. The term 'asymptotic regression' would be preferable.

Ex Post. 'From afterwards.' The term generally means a retrospective view of an event or set of events. Actual observed results may be compared with those predicted by a given model, and the resultant comparison is ex post. Any modification of the model or hypothesis is also termed ex post. In financial statistics the term is sometimes employed for modified budget estimates, taking into account exceptional events which have occurred since the original ex ante estimates were made.

Extensive Sampling. A term which is used to denote sampling where the subject matter, or geographical coverage of a sample has been diffuse or widespread as opposed to intensive, where it gets narrowed to a small field. Extensive sampling may refer either to a case where a wide variety of topics have been covered superficially, rather than a few topics in detail or a large area has been surveyed broadly, rather than a small area studied in detail. The term might also be used with reference to time, that is to say, of sampling covering a long period.

Extremal Quotient. The relative difference between the largest and smallest values in a smample, as the range is the absolute difference.

External Variance. An obsolete term which was introduced by Schultz (1930). This term is used to denote the variance of predictions of the future movements of a particular time series which is based upon a given form for its trend. The components contributing to the variance have been errors in the estimation of the parameters, not those arising from an incorrect choice of trend line or from superposed random variation.

Extrapolation. The extension of a graph beyond the upper or lower limit of the observed range of the basic statistics, or the calculation from a regression equation of an estimated value or values of the dependent variable corresponding to a value or values of one or more of the independent variables outside the observed range. It is a process which theoretical statisticians frown upon, and indeed, it needs using with case. The further the graph is extended beyond the observed range the more dangerous the process becomes ; the same applies to extrapolated values calculated from a regression equation.

Extremal Statistic. Refers to any function which depends upon observations at both extremes of the sample. Examples of external statistic include the range and the extremal quotient.

Extreme Mean. Refers to one of a set of mean values, *e g.*, in the analysis of variance which lies at the extreme of those values, *i.e.*, the largest or the smallest.

Extreme Values. Refers to the largest or smallest variate values which are borne by the members of a set. Slightly more generaly, the expression signifies values neighbouring the end-values.

Extreme Value Distributions. Refer to the distribution of the largest (smallest) observation in a sample. The distributions may be exact or asymptotic in form and there have been three types in accordance with the population from which the sample has been drawn.

F

Factor. This word finds use in statistical contexts in many senses :

(*a*) in the ordinary mathematical sense, *e.g.*, it denotes a factor of an algebraic expression.

(*b*) it is also used to denote a quantity under examination in an experiment as a possible cause of variation, *e.g.*, in a 'factorial' experiment ;

(*c*) (adapted from psychology) in multivariate analysis, it is used to denote a function of the observed variates, usually linear, which may be regarded as part of those variates ; and hence as a 'factor' of the variation ;

(*d*) it is used to denote a constituent item in an average or index number.

Factor Analysis. A statistical procedure used to determine the basic irreducible variables (factors) underlying a large number of interrelated variables. When measurements of a large number of variables have been obtained (each representing, for example, a test or a test item), factor analysis may be used to reduce them or a smaller number of basic types or factors. Factor analysis, in essence, involves the analysis of the correlations among the variables.

Factorial. This term refers to the product of all the positive integers from 1 up to and including a given integer. Symbol : $n\,!$ where n is the given integer. Thus 5 factorial, written 5 ! ; meanes

$$5\times4\times3\times2\times1\ (=120)$$

Factorials find use in working out probabilities in the poisson distribution and binomial distribution, in permutations and combinations, etc.

Factorials very quickly become large numbers. By convention, 0 ! is given the value 1.

1 ! = 1
2 ! = 2
3 ! = 6
4 ! = 24
5 ! = 120
6 ! = 720
7 ! = 5,040
8 ! = 40,320
9 ! = 362,880
10 ! = 3,628,800

For large values of n, n ! may be given approximately by

$$n\,! \simeq \sqrt{2\pi n}\, n^n e^{-n}, \qquad \text{or } n\,! \simeq \sqrt{2\pi}\; n^{(n+\frac{1}{2})}\, e^{-n}$$

which can be rewritten as *Stirling's approximation to n* !

$$\log_{10} n\,! = .39909 + (n+\tfrac{1}{2}) \log_{10} n - .4342945\, n$$

The concept is important in such cases as :

1. calculating probability in sampling without replacement :
2. calculating expected values in the poisson distribution ;
3. computing probability in fisher's exact probability test.

The term should not be confused with factor.

Factorial Distribution. A discrete distribution for which the successive frequencies have been factorial qualities. Irwin (1963) defined a whole class of such distribution of form

$$f(r) = \frac{ka_r}{\theta(\theta+1)\ldots(\theta+r)}\ \theta > 0,\ r = 0, 1, 2, \ldots$$

Factorial Experiment. An experiment which has been designed to examine the effect of one or more factors, each factor being applied at two levels at least so that differential effects might be observed. The term finds frequent use in a slightly narrower sense, as describing an experiment investigating all possible treatment combinations which may be formed from the factors under investigation. The 'level' of a factor denotes the intensity with which it is brought to bear. It could be measured quantitatively, as when fertiliser is applied to plots in a given weight per unit area, or qualitatively, as when patients are considered at two levels 'inoculated' and 'not inoculated'. It gives rise to calculations for analysis of variance.

Factorial Sum. Refers to the sums which are entering into the calculation of factorial moments. If the frequency of a variate value $r(r=0, 1,.., k)$ has been f_r the factorial sum of order j would be

$$\sum_{r=0}^{k} \{f_r r(r-1)...(r-j+1)\}.$$

Factor Rotation. The final stage of factor analysis has been in geometric terms, a rotation of the factor axes so as to achieve some correspondence between the numerically derived factors and explanatory entities; that is to say to seek a structure. Analytic methods could be divided broadly into those dealing with orthogonal factors, where the loadings get uncorrelated, and oblique factors involving correlated loadings. One group of methods pursues the maximisation of the scatter among factor loadings; for example, quartimax and varimax for orthogonal rotation and oblimax for oblique rotation. An alternative approach, using the minimisation of cross-products, is called quartimin and covarimin, which latter method is the inverse of varimax.

Fair Game. In the theory of games, a game consisting of a sequence of trials has been deemed to be a 'fair' game if the cost of each trial is equal to the expected value of the gain from each trial. A 'fair' game in this sense may not be fair as between a pair of adversaries with unequal resources: it is well known that at a 'fair' game the player with the larger sum to stake has the better chance of ruining his opponent.

F-Distribution. See variance ratio distribution.

Fermi-Dirac Statistics. See Bose-Einstein statistics.

Fertility Rate. Refers to the number of live births in a unit period which is expressed, as a proportion (usually per thousand) of potentially fertile women in the population concerned. 'Potentially fertile' is generally defined by reference to age, *e.g.*, by taking women from 15-50 years old. This crude or general fertility rate may be refined or standardised in respect of age, social class, etc.

Fertility rate is usually expressed per thousand per year.

Fiducial Distribution. Refers to a distribution of a parameter needed for fiducial inference about that parameter. It is not a probability distribution in the customary sense, but gets derived from the distribution of estimators containing all the relevant information in the sample. In earlier literature it was sometimes referred to as a 'fiducial probability distribution.

Fiducial Inference. A type of statistical inference which is based on fiducial distribution. It was introduced by R.A. Fisher (1930). The object of the inference has been to make probabilistic statements about the values of unknown parameters and to that extent it resembles the theory of confidence intervals.

Fiducial Limits. Sometimes called *fiduciary limits* or *confidence limits* ; the limits within which a derived figure such as an average or a standard deviation is true of the universe. The fiducial probability that the true average of the universe falls within the fiducial limits set by the standard deviation of the statistical series applied to the average of the series, is 68%. One-and-a-half times the standard deviation has a fiducial probability of 87%, and twice the standard deviation has one of 95%. If the average of a series is 156, and the standard deviation is 20, then the fiducial probability is 68% that the true mean of the universe falls within the fiducial limits of 156 ± 20 *i.e.*, between 136 and 176.

Fiducial Probability. See fiducial distribution.

Finite Markov Chain. A Markov chain $\{x_n\}$ is termed as a finite chain with k states when the number of possible values of the random variables $\{x_n\}$ is finite and equal to k.

Finite Multiplier/Finite Population Multiplier. A term which is synonymous with finite population correction and is not now widely used although it was at one time considered a preferable term. If a sample of n values could be drawn without replacement from a population of limited size N, the sampling variances of the statistics derivable from it have been found to depend, in general, on N as well as n. For example the variance of the sample mean $\bar{x}$ may be written

$$\text{var}\,\bar{x} = \frac{\sigma^2}{n}\left(1 - \frac{n}{N}\right)$$

$$\text{where } \sigma^2 = \frac{1}{N-1}\sum_{i=1}^{N}(x_i - \mu)^2 \text{ and } \mu = \frac{1}{N}\sum_{i=1}^{N} x_i.$$

The factor $1 - n/N$ is sometimes termed as the finite multiplier or the finite sampling correction.

Finite Population. Refers to a population of individuals which have been finite in number.

Finite Population Correction. A correction factor which is applied to the expression for the variance of the means of samples drawn from an infinite parent population, to allow for it being of finite size. When samples of size *n* have been taken from an infinite parent population with standard deviation σ (or from a finite one with replacement) the variance and standard deviation of the sample means would given by :

$$\text{Variance} = \frac{\sigma^2}{n} \qquad \text{SD} = \frac{\sigma}{\sqrt{n}}$$

If, however, the samples have been taken from a population of size N, without replacement, the variance and standard deviation of the sample means are given by :

$$\text{Variance} = \frac{\sigma^2}{n}\left(\frac{N-n}{N-1}\right) \qquad \text{SD} = \frac{\sigma}{\sqrt{n}}\sqrt{\frac{N-n}{N-1}}$$

The difference between these two expressions for the variance has been the factor

$$\frac{N-n}{N-1}$$

which is the finite population correction.

If N is large in comparison with *n*, the finite population correction may be ignored.

Finite Sampling Correction. Refers to as alternative term for the finite multiplier. It has been thought as objectionable on the grounds that it has been not a 'correction' to an inexact result.

First Limit Theorem. If a sequence of distribution functions is tending to a single distribution function F then the corrosponding characteristic functions would tend uniformly in any finite interval to the characteristic function of F. This is generally termed as the First Limit Theorem. Although it is known to earlier writers, it seems to have been proved rigorously for the first time independently by *p.* Levy and Cramer about 1925.

Fisher's 'B' Distribution. Refers to the distribution of the square root of a non-central chi-squared variate. It has been a hypergeometric function and was derived by Fisher in a study of the multiple correlation coefficient.

Fisher's Ideal Index Number. Refers to the geometric mean of Laspeyres's and Paasche's index numbers. Laspeyres and Paasche are biased, if at all in opposite directions ; where the index is one of prices, the former usually upwards and the latter downwards. Professor Irving Fisher, the American economist and author of the standard work on index numbers, *The Making of Index Numbers,* suggested that a cross of Laspeyres and Paasche would therefore give the ideal index number. The subject is more fully discussed under index numbers.

Fisher Model. A term which was proposed by Ogawa (1963) to denote the general class of experiment designs without technical errors but based upon Fisher's randomisation procedure.

Fisher's Distribution. Refers to the distribution of the ratio of the variances of two independent samples from a Normal population or, a little more generally, to the ratio of two independent quantities, each of which gets distributed as χ^2.

Fisher Exact Probability Test. A non-parametric statistical technique used for analyzing discreate data from two samples that are small in size. The Fisher test is used to determine whether there is a significant difference between two samples in the proportion of cases that fall in each of two mutually exclusive categories. The Fisher test is used in preference to the chi-square test when the number of cases is small and the data are in a fourfold (2×2) table.

It is a method of testing association in (2×2) contingency tables where the total number of observations is less than fifty, and where, in consequency, Yates modification of the chi-squared test is unreliable.

The value of P is obtained by using the formula

$$P=\frac{(a+b)!(c+d)!(a+c)!(b+d)!}{n!a!b!c!d!}$$

where a, b, c and d are the cell frequencies and n is the total number of observations (*i.e.*, $n=a+b+c+d$). The null hypothesis is that of non-association between row and column categories. If the value of P falls below 5 per cent or 1 per cent. The null hypothesis is rejected at the 5 per cent or 1 per cent levels respectively.

Fisher's Z-transformation (Of the Correlation Coefficient). Refers to a transformation of the correlation coefficient r according to the formula

$$z=\tan h^{-1}{}_r.$$

The distribution of z for samples from a bivariate Normal population approaches normality of form much more rapidly than does that of r; and its variance, unlike that of r, has been little affected by the population correlation coefficient even for samples of moderate size. It has been therefore useful in many problems involving estimation and tests of significance.

Fisher's z transformation has been used :

(*a*) In testing the hypothesis that the correlation coefficient of the parent population of a bivariate distribution has a particular value :

(*b*) In setting up confidence limits for an estimated value of the parent population's correlation coefficient.

When a correlation coefficient r is calculated, the answer obtained is based on a sample. If, for example, 20 Siames cats were used in a dietary experiment, the correlation coefficient between their masses (in kg) and daily food intake (in calories) could be calculated, giving a value of, say, 0.6. The 20 cats are a sample from a very large parent population, all Siames cats. Thus the value of 0.6 is only an estimate of the parent population's correlation coefficient, based on a sample of size 20.

When significance level tables has been used for interpreting correlation coefficients, the null hypothesis being tested is : There has been no correlation. The sample has been drawn from a parent population with correlation coefficient 0.

The alternative hypothesis has been that the parent correlation coefficient is not zero. Thus a two tail test is involved (see one-and two-tail tests). In order to test this hypothesis, it should be necessary to know the distribution of correlation cofficients for samples of size 20. However, Fisher's transformation overcomes this problem by converting this distribution into a normal one. If r has been a particular sample correlation coefficient, the corresponding value of z would be as follows :

$$z=\frac{1}{2}\, ln \left(\frac{1+r}{1-r}\right)$$

The distribution of z has been normal, with :

$$Mean=z_0$$

$$Standard\ deviation=\frac{1}{\sqrt{n-3}}$$

Thus, if ρ_0 has been the true parent correlation coefficient, z_0 (the mean of the transformed distribution) has been found from the equation above as :

$$z_0 = \frac{1}{2}\, ln\left(\frac{1+\rho_0}{1-\rho_0}\right)$$

In the case of the 20 Siamese cats, the null hypothesis has been that $\rho_0 = 0$, giving $z_0 = 0$. The z value (using $r = .6$) is

$$\frac{1}{2}\, ln\left(\frac{1+6}{1-6}\right) = .693$$

This calculation of the z value can be avoided by using the Fisher z-transformation.

The standard deviation σ for $n = 20$ is :

$$\frac{1}{\sqrt{20-3}} = .243$$

If we apply the normal distribution to these data (Fig. 1), the shaded areas are each 2 $(1' - \theta\ (2{\cdot}85)) = .0044$. So, if the null hypothesis, that $\rho_0 = 0$, has been true, the probability of getting a value of r at least as far away from that value as the one observed (*i.e.*, $r \geqslant .6$, or $r \leqslant -.6$) is .0044, the significance level of the result. There has been thus good reason to reject the null hypothesis of no correlation.

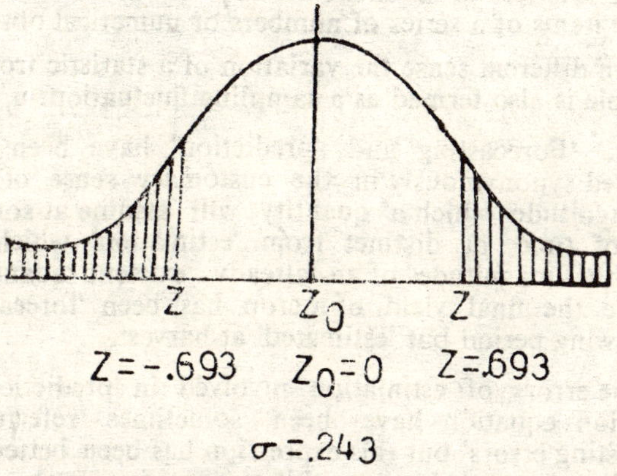

Fig. 1. Fisher's z-transformation. The transformed distribution. Standardizing the value of z, *i.e.*, .693, gives :

$$\frac{.693-0}{.243} = 2.85$$

This answer, that the correlation coefficient has been significant at better than the .5% significance level, could have been obtained without the use of Fisher's transformation.

Fixed Base Index. Refers to an index number for which the base period for the calculations has been selected and remains unchanged during the lifetime of the index. It has been in contradistinction to a chain base index.

Fixed-Base Index Number. A series of index numbers with a common base (=100). The base usually covers one of the periods, times or places within the series, not necessarily the first in a time series. It serves as a common standard of comparison. The weighting system does not necessarily follow the base. The pattern of weights applies to the base period or place when Laspeyres's index is used, and to the given period or place when Paasche's index is used. A chain-base index applicable only to time series, uses the weights either of the given period or of the preceding period, the base always being the preceding period.

Fixed Effects (Constants) Model. Refers to an alternative designation for model I (or first kind) in analysis of variance.

Fixed Sample. If a survey is repeated on several occasions, but observations are taken on the same sample instead of a new sample for each occasion, the sample is considered to be fixed.

Fluctuation. Refers to a movement up or down between consecutive items of a series of numbers or numerical observations.

In a different sense the variation of a statistic from sample to sample is also termed as a sampling fluctuation.

Forecasting. 'Forecasting' and 'prediction' have been generally employed synonymously in the customary sense of assessing the magnitude which a quantity will assume at some future point of time : as distinct from 'estimation' which tries to assess the magnitude of an already existent quantity. For example, the final yield of a crop has been 'forecast' during the growing period but 'estimated' at harvest.

The errors of estimation involved in prediction from a regression equation have been sometimes referred to as 'forecasting errors' but this expression has been better avoided in such a restricted sense. Likewise terms such as 'index numbers of forecasting efficiency', in the sense of residual error variances in regression analysis have to be avoided.

Fourfold Table. A statistical tables with four cells, or places for the entry of data. A fourfold table is formed by the cross-tabulation of two variables each of which has just two categories, so that there are four possible combinations of categories. For example, if the variable sex, with the two categories men and women, and the variable schooling, divided into just two categories—graduated from high school and did not graduate from high school—were cross-tabulated, there would be four possible combinations of categories : men who graduated from high school, men who did not graduate from high school, women who graduated from high school, and women who did not graduate from high school.

Fourier Analysis. Refers to the theory of representing functions of a variable t as the sum of a series of sine and cosine terms of type $a_j \cos (2\pi_j/\lambda_j)$, $j=0, 1,...$ The λ's has been not necessarily commensurable and hence the analysis has been more general than harmonic analysis which considers series of terms such as $\cos (2\pi_j/\lambda)$ where λ has been some constant.

Frame. Refers to a representation of the items which are available to be chosen for a sample. This could be a telephone directory, a map, an electoral register, a list of the cattle in a herd, etc.

Frechet Distribution. The Frechet distribution has been alternative name for the type III, or second asymptotic, distribution of extreme values. The distribution function has been

$$F(x)=exp(-y^{-k}) \text{ where } y=(x-a)/b \,;\, b>0 \text{ and } x\geqq a$$

and k a positive constant related to the shape.

Freehand Method. Refers to a method of describing the relationship in a series of data, ordered in time or space, whereby the general trend has been estimated by drawing a line freehand through or near the series of plotted observations.

Frequency. The number of occurrences of :

1. a particular event ; or
2. a particular class or interval group ; or
3. particular values of a variate.

For example, if in a degree examination there are three distinction results, twelve passes with credit, thirty ordinary passes and two failures, then the frequency of distinctions in the examination is three, etc.

Example : in 20 shots at a target, a marksman makes the following scores :

4	3	2	4	5	2	4	2	5	1
0	5	1	5	2	5	5	0	1	4

The frequencies of the different scores are thus :

Score	0	1	2	3	4	5
Frequency	2	3	4	1	4	6

Frequency Curve. Refers to the graphical representation of a continuous frequency distribution, the variate being the abscissa and the frequency the ordinate. The frequency curve could be viewed as the limiting form of the frequency polygon as the number of observations tends to become infinitely large, and the class intervals indefinitely small.

Thus frequency curves are used where class intervals become very small and numbers of observations are very large. The ordinate axis is used to measure frequency values and the abscissa is employed to measure variate values.

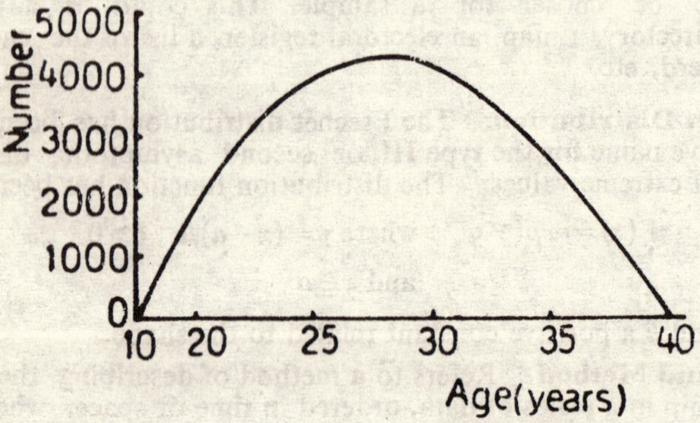

Numbers and ages of people taking a given professional examination.

Frequency Distribution. A classification of data showing the number of occurrences of each class interval (subdivision of a quantitative variable) or category of a variable. In the case of a quantitative variable the class intervals are arranged in order of magnitude. A frequency distribution provides a basic ordering of data, and is usually the first step in statistical analysis.

Frequency Distribution Bimodal. A frequency distribution having two modes or peaks. In a bimodal frequency distribution there are two separated (noncontiguous) class intervals that have the largest number of cases in the distribution of a variable. When presented graphically, the curves has two humps. If the two modes are widely separated an arithmetic mean would have little significance. See class interval; frequency distribution curve; mean, arithmetic; mode.

Frequency Distribution, Cumulative. A frequency distribution that presents for each class interval (subdivision of a quantitative variable) the total number of cases in that interval and below (or above) it. Thus the number of cases given for each class interval includes the cases that fall in that interval and in all precefiing intervals. See variable.

Frequency Distribution, Multimodal. A frequency distribution with three or more modes.

Frequency Distribution, Poisson. See Poisson distribution.

Frequency Distribution, Probability. See probability distribution.

Frequency Distribution, Single-Humped. Sce frequency distribution unimodal.

Frequency Distribution, Symmetrical. A frequency distribution in which the distribution of cases is identical on both sides of the mode (most frequent value, or class interval with the greatest number of cases). In a symmetrical frequency distribution the mode, the median, and the arithmetic mean coincide.

Frequency Distribution, Unimodal. A frequency distribution that has one mode, or peak of frequencies.

Frequency Distribution, Curve. The graphic presentation of a frequency distribution. The frequencies (number of cases) are listed along the vertical axis and the class intervals (divisions of the variable) or their midpoints are listed along tee horizontal axis. The frequency of each class interval is then plotted, and a line is drawn connecting these points. A frequency distribution curve is often referred to as a frequency polygon.

Frequency Distribution Curve, Leptokurtic. A frequency distribution curve having a pronounced peak, or a higher degree of kurtosis (peakedness) than a normal frequency distribution curve.

Frequency Distribution Curve, Mesokurtic. A frequency distribution curve that is neither peaked nor flattened, but is moderately rounded. A normal frequency distribution curve is a mesokurtic curve.

Frequency Distribution Curve, Normal. A frequency distribution curve in which the highest frequencies are in the center with steadily decreasing frequencies to the right and left of the central value. A normal curve is unimodal, symmetrical, and bell shaped.

Frequency Distribution Curve, Platykurtic. A frequency distribution curve that is quite flattened. A platykurtic curve has less kurtosis (peakedness) than a normal frequency distribution curve.

Frequency Distribution Curve, Skewed. A frequency distribution curve in which the highest frequencies are to the right or left of the center. The curve is asymmetrical and the range of cases is greater on one side of the mode (the highest point or class interval with the greatest number of cases) than on the other.

Frequency Function. It is an expression which is giving the frequency of a variate value x as a function of x; or, for continuous variates, the frequency in an elemental range dx. Unless the contrary gets specified the total frequency could be taken to be unity, so that the frequency function represents the proportion of variate values x. From a more sophisticated standpoint the frequency function is most conveniently considered as the derivative of the distribution function. The generalisation to more than one variate has been immediate.

Frequency Polygon. A graphical depiction of a discontinuous (or discrete) frequency distribution, of which the frequency curve is a limiting form. Variate values get plotted against the abscissa and frequencies against the ordinate axis. When it has been used for discrete value categories, the frequency polygon is constructed by plotting frequencies for each variate value and joining the points together in the manner shown in the diagram overleaf.

Frequency Surface. A three-dimensional representation of a scatter diagram. The area of the graph is divided into a grid, each part of which is raised to a height proportional to the frequency of the points lying within it. It is the bivariate analogue of the frequency curve.

Frequency Table, A frequency distribution presented in the form of a statistical table. The class intervals, or subdivisions, of a quantitative variable are listed from lowest to highest. The cells of the table indicate the number of occurrences of each class interval or category in a set of data. Sometimes the original class intervals used in a study are grouped into a smaller number for convenience in tabuar presentation.

Frequency Theory of Probability. This theory regards the probability of an event as the limit of the frequency of occurrence of that event in a series of *n* trials as *n* tends to infinity. The existence of this limit has been an axiom of the theory as proposed by von Mises (1919), but later axomatisations (*e.g.*, by Kolmogorov, 1933) have been avoiding the difficulties associated with it by taking the probability as a measure associated with a set of points (events) and proceeding on the basis of measure theory. This avoids the difficulty only for a mathematician. For the statistician the problem of relating probability to frequency of occurrence remains.

Friedman's Test. Essentially it is the same as pitman's test for homogeneity of means in otherwise identical populations, except that observations are replaced by ranks prior to application of the test (Friedman, 1937). This test is an adaptation of the chi-squared test to ranked measurements to compare three or more matched samples. It was devised by Friedman in 1937, and involves obtaining a chi-squared statistic by squaring rank totals instead of summing the squares of differences between observations and expected values. The sum of squares of ranks is obtained (R) and the formula applied is :

$$\chi_r^2 = \frac{12R}{Nk(k+1)} - 3N(k+1)$$

where *k* denotes the number of observations, N the number of values of each observation, and χ_r^2 a symbol indicating the rank chisquared coefficient.

Friedman χ_r^2 Test. A non-parametric statistical test that provides an analysis of variance (dispersion of values) based on ranks, or relative position. It is used to test whether the obtained differences between two or more groups (matched with respect to all relevant variables, including number of members, other than the independent variable) may be considered statistically significant or whether they are probably due only to chance. This test may also be used to compare the same group under different conditions.

Friedman's Two-way Analysis of Variance by Rank. Refers to a non-parametric test of the null hypothesis that several matched samples have been drawn from the same parent population.

F-Test. A test of whether there is a statistically significant difference between two or more groups, or whether the groups are probably drawn from identical populations (or the same population) with differences between them due only to chance. The F test is based on the analysis of variance or the dispersion of values, and involves the relationship of the variance between groups to the variance within the groups. It is a parametric test, requiring the usual parametric assumptions (primarily concerning the population distribution in the population from which the samples were drawn).

Full Information Method. In econometrics, it refers to a method of deriving estimates of parameters in a stochastic model which have been subject to all the *a priori* restrictions of that model.

Function. In statistics used in the mathematical sense, that is, to quote the Oxford Dictionary, 'variable quantity in relation to other (*s*) in terms of which it may by expressed or on which its value depends. Mathematicians and mathematical statisticians use a form of shorthand, *e.g.*,

$$Y=f(X)$$

which simply means that Y is a function of X. In practical statistics, an equation like this serves merely as a preliminery model : it saves writing out a verbal model. Before it can be applied in practice, it has to be converted to an equation which indicates the type of relationship that exists theoretically, or empirically, between the variable quantity and the other (*s*), *e.g.*,

$$Y=aX+b$$

$$Y=bX^a$$

$$Y=aX^2+bX+C$$

$$Y=ba^{\chi}$$

Function Central Limit Theorem. An alternative name for Donsker's theorem.

Fundamental Probability Set. Refers to a set of objects or events which are basic to a probabilistic situation, in the sense that all other objects or events under consideration are derived from them by compounding. It is evident that all probabilities are expressible by the rules of addition, multiplication, etc. of probabilities in terms of the probnbilities of the fundamental set. Failure to specify the fundamental set explicitly leads to confusion, occasionally to error, and in particular to a number of paradoxes. The fundamental probability set has been sometimes called the reference set.

Fundamental Random Process. See Brownian motion process.

Future Value. Terminal valuation.

G

Gabriel's Test. Refers to a simultaneous comparison procedure for mean values (1964) in analysis of variance. It has been based upon the 'between-groups' sum of squares for the $(2^k - k - 1)$ subsets of two or more mean values of the k groups and has been tested against the fixed critical value of the F-distribution with $(k-1, n-k)$ degrees of freedom. The sample sizes (n_i) need not be equal.

Galton Ogive. Under certain conditions, *e g.*, when a frequency function has been unimodal, the distribution curve resembles a letter 'S'—the ogive form, a term which was borrowed by Galton (1875) from architecture and used particularly for the distribution curve of the Normal distribution.

Galton-McAllister Distribution. Refers to a frequency distribution in which the logarithm of the variate, or some simple linear function of it Normally gets distributed. The distribution was suggested by McAllister (1879). It is now more generally termed as the lognormal distribution.

Gambler's Ruin. Refers to the name which is assigned to one of the classical topics in probability theory. A game of chance can get related to a series of Bernoulli trials at which a gambler wins a certain predetermined sum of money for every success and losses a second sum of money for every failure. The play may take place until his initial capital gets exhausted and he gets ruined. The statistical problems involved get concerned with the probability of the ruin of a player, given the stakes, initial capital and chances of success, and with such matters as the distribution of the length of play.

There have been many variations to this classical problem, which gets closely associated with problems of the random walk ; in particular, of sequential sampling.

Game Theory. A game is a conflict situation and *game theory* is concerned with devising and giving effect to the strategies of such situations, namely the main features of a plan prepared before action commences and including an objective.

Other related terms have been as follows :

Pay-off—the value of the game to a player.

Zero-sum game—game in which all the pay-offs for all prayers total zero, *i.e.*, what one player gains, other lose.

Non-zero-sum game—game in which all the pay-offs for all players do not total zero.

Finite game—game in which each player has a finite number of possible strategies.

Infinite game—game in which at least one player has an infinite number of strategies.

A *2, 3, 4...n-person game*—game with 2, 3, 3...*n* players or conflicting sets of interests.

Game theory is an important branch of operational research and is related to decision theory and operational gaming. It is also linked closely with linear programming in that linear programmes can often be used to obtain the best strategies to use in a conflict.

Game theory deals with situtations of conflict between two or more opposed sets of interests and is used to calculate the optimum strategy which maximizes the winnings or minimizes the losses of one or more of the 'players'. (The usual criterion is to minimize the maximum losses).

The working procedure consists of six steps as follows :

(*i*) Formulate the problem.

(*ii*) Specify the pay-off criterion.

(*iii*) Set out the pay-offs in the form of a 'game matrix'.

(*iv*) Carry out the requisite arthmetic operations until the solution is obtained.

(*v*) Test the solution.

(*vi*) Implement.

Gamma Coefficients. Refers to the ratios, analogous to moment-ratios or the beta coefficients, which have been based upon the cumulants :

$$\gamma_1 = k_3/k_2^{3/2},\ \gamma_2 = k_4/k_2^2$$

and, in general

$$\gamma_r = k_r/k_2^{r/2}$$

Gamma Distribution. A particular kind of distribution, also known as type III in the Pearsonian system of frequency distributions. It is continously distributed from zero of infinity and is also unimodel. The formula for the frequency distribution with probability density function $f(x)$ is given by :

$$f(x) = \frac{x^{a-1} e^{-\lambda x}}{\lambda^{-a} \lambda(a)}$$

where a and λ have been parameters of the distribution, and $\lambda(a)$ denotes the gamma function, given by :

$$\lambda(a) = \int_0^\infty x^{a-1} e^{-x} dx \text{ (for } a \text{ real and positive).}$$

If a has been a positive integer, $\lambda(a) = (a-1)!$

The gamma distribution has

$$Mean = \frac{a}{\lambda}$$

$$Variance = \frac{a}{\lambda^2}$$

The graph of the gamma distribution varies in shape according to the value of a, as shown in figure.

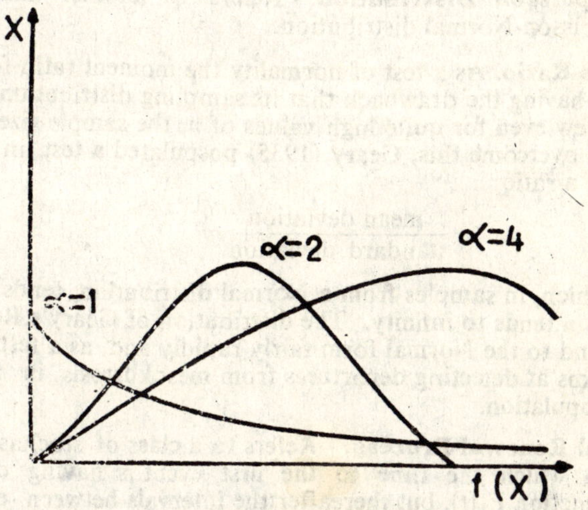

Gamma distribution. The different curves for values 1, 2, 4 of the parameter a.

In the case $a=1$, the gamma distribution has been the *exponential distribution.*

Gantt Progress Chart. Refers to on application of the bar chart due to Gantt, of use in industrial statistics. An actual performance or output has been expressed as a percentage of a quota or planned performance per unit of time. Account may also be taken of the **cumulative** performance by plotting it, with the planned cumulative performance, as ordinate against time as abscissa.

Gauss Distribution. Refers to an alternative name for the Normal distribution.

Gaussian Curve. See normal curve.

Gauss-Markov Theorem. Refers to a fundamental theorem which deals with an unbiassed estimator of a population characteristic, and is based upon a linear combination of sample observations drawn from that population. The theorem has been to the general effect that an unbiassed linear estimator of a parameter has been 'best *i.e.*, has minimum variance, when the estimator has been obtained by least squares. The theorem could be extended in many directions, *e.g.*, by regarding the simultaneous estimation of several parameters or linear functions of them.

Gauss-poisson Distribution. Refers to another name for the Poisson-Normal distribution.

Geary's Ratio. As a test of normality the moment ratio for kurtosis is having the drawback that its sampling distribution has been skew even for quite high values of n, the sample size. In order to overcome this, Geary (1935) pospulated a test in the form of a ratio

$$\frac{\text{mean deviation}}{\text{standard deviation}}$$

which, in samples from a Normal distribution, tends to $\sqrt{(2/\pi)}$ as n tends to infinity. The distribution of Geary's Ratio would tend to the Normal form fairly rapidly and, as a test, the ratio aims at detecting departures from mesokurtosis in the parent population.

General Renewal Process. Refers to a class of stochastic process in which the time to the first event is having distribution function $F_1(t)$, but thereafter the intervals between events have been independently and indentically distributed with distribution function $F(t)$. $F(t)$ has been termed as the 'renewal distribution' whilst $F_1(t)$ has been referred to as the distribution of the first renewal time.

The term renewal process has been used if $F_1(t) \equiv F(t)$.

Generalised Binomial Distribution. Refers to an alternative name for the Poisson binomial distribution.

Generalised Bivariate Exponential Distribution. Refers to a generalisation of the bivariate form which is obtained directly from the multivariate exponential distribution. This generalisation (Marshall & Olkin, 1967), has been important in connection with the form of the underlying model used in life assessment problems.

Generalised Distribution. This expression is found in two senses, (*i*) as a more complicated form similar to some known distribution ; (*ii*) as obtained from a known distribution when its parameters are themselves random variables. The latter usage should get discarded in favour of compound distribution.

Generalised Maximum Likelihood Estimator. Refers to an approach to parameter estimation which is involving the concept of asymptotic efficiency proposed by Weiss & Wolfowitz (1966) to remove some of the problems associated with classical maximum likelihood estimators.

Generalised Multinomial Distribution. Refers to an n-dimensional distribution of discrete variates which were proposed by Tullis (1962) which includes an additional parameter ρ as the correlation coefficient between pairs of variates with common mean and marginal distributions which have been multinomial.

Generalised Normal Distribution. A little used alternative name for Kapteyn's univariate distribution which is not to be confused with the multivariate Normal distribution.

Generating Function. Refers to a function of a parameter t which, when expanded as a power series in t, gives as the coefficients the values of some quantity of statistical interest such as the probability of events or the moments of a frequency distribution. The characteristics function has been an important case of a moment generating function, a phrase which has been often abbreviated to *m.g.f.* The theory of probability is making use of the generating function approach since the time of De Moivre in the first half of the 18th century.

Geometric Distribution. Refers to a distribution in which the frequencies fall off in geometric progression as the variate values increase. The expression usually gets confined to discontinuous distributions and does not, for example, include the exponential distribution.

The continuous form of geometric distribution is sometimes called an exponential distribution.

The probability distribution $p(x)$ is given as follows :

$$p(x)=pq^{x-1} \quad \text{where} \quad 0<p<1, q=1-p$$

It represents the probability of the first success in a series of bernoulli trials occurring at the xth attempt. Each trial has probability p of success, probability q of $1-p$ of failure.

For this distribution

$$Mean=\frac{1}{p}$$

$$Variance=\frac{q}{p^2}$$

The distribution should not be confused with that which appears normal when plotted on a horizontal logarithmic axis, of which the mode and median are the geometric mean.

Ceometric Mean. Refers to the average value of a set of n numbers, which is expressed os the nth root of their product. Thus, the geometric mean of n numbers $x_1, x_2, \ldots x_n$ is as follows :

$$\sqrt[n]{x_1\, x_2 \ldots x_n}$$

Geometric Moving Average. Refers to an expression to be avoided, as it has been not based on geometric means.

Geometric Progression. A series (or sequence) of numbers ranked in rising or falling order such that the ratios between successive pairs of terms are constant.

Example : The series 1, 3, 9, 27, 81......is a geometric progression because the ratio between each successive pair (1, 3), (3, 9), (9, 27) is constant (=3). The median of the series is the geometric mean (or geometric average). The sum of a geometric progression is

$$a\,\frac{r^n-1}{r-1}$$

where a is the first term of the series, r is the ratio between each successive pair of terms and n is the number of terms. Thus, in the above example, $a=1$, $r=3$ and $n=5$, so that the sum of the terms is

$$1\times\frac{3^5-1}{3-1}=\frac{242}{2}$$
$$=121$$

Among other things; the concept of geometric progression is important in compound interest calculations for the computation of sinking-fund values, where the increment of a rate of interest (*r*) may be expressed as $(1+i)$, *i* being the rate of interest, and therefore the sum of a sinking fund of *n* terms (*i.e.*, $n-1$ years) may be calculated using the formula

$$\frac{(1+i)^n-1}{i}$$

Geometric Range. This expression is found very infrequently in two senses; (*i*) as the ratio of the extremes of a sample, (*ii*) as the geometric mean of the extremes of a sample. Neither usage is having much to recommend it.

Given Period. In the construction of index numbers it becomes necessary to relate data at two points of time, a base period and some other period which may be earlier or later in time, the 'given' period.

Gompertz Curve (Growth Curve). Refers to a trend curve which is representing either growth or decline where the rate of change either increases or declines at a constant percentage or rate. The equation of the curve has been

$$Y=ae^{-bt}$$

where Y denotes the dependent variable, *a* denotes a constant and *bt* denotes the constantly increasing or declining rate of growth.

Goodman-Kruskal Tau. Refers to a measure of association for cross classification in contingency tables. It has been asymmetric between the two qualities concerned, being based on a probability interpretation involving the nation of predicting one given the other.

Goodness of Fit. A measure of the agreement between observed and expected values. The expected values may be determined by a regression model (in the case of regression analysis) or some other proposed or theoretical model. The extent of goodness of fit is measured by the extent to which expected values plotted on a given line or curve do not differ from observed values. Generally the lack of fit is measured in terme of some function of the sum of squared differences between observed statistics and expected values. The chi-squared test can be used to test goodness of fit in many cases.

The term is used especially in relation to the fitting of theoretical distributions to observation and the fitting of regression lines.

Goodness of fit may be measured, for grouped data, by the statistic χ^2 (see chi-squared test), where :

$$\chi^9 = \sum_{\text{all groups}} \frac{(O-E)^2}{E}$$

for the correct number of degrees of freedom, where O is the observed number in a group, and E the expected number.

Grade. Galton used the word to mean the relative cumulative frequency of the variate values or the proportion of total frequency having values less than or equal to the variate value. In the table under cumulative frequency, for example, 70 of the 250 office workers have salaries of Rs. 1999 or under. The grade of the value Rs. 1999 is therefore 7/25ths, or 0.28.

If the categories used in the cumulative frequency distribution are discrete, the calculation of grade is similar except that only half the frequency value of the highest variate class is included. Thus if total frequency is 1000 and variate ranges from 1 to 10, and if the cumulative frequency of variate values from 1 to 6 is 575 and the frequency of variate value 7 is 80, the calculation of grade for variate value 7 would be

$$\frac{575+\frac{1}{2}(80)}{1000}=0.615$$

Grand Average. The average of a universe as distinct from that of a sample of the universe.

Grand Mean. A mean of a number of mean values. The term is used, *inter alia*, in analysis of variance and in mean range method of computing standard deviation. Thus, in the latter usage, a group of items is divided into subsets of a specified size (*e.g.*, $n=6$). Then the means and ranges of each of the sets are calculated. The mean of the ranges is determined and multiplied by a factor whose value depends on the number of items used in each subset. The result is an approximation of standard deviation. The parameter mean may be estimated by computing the mean of each of the means of the subsets. This mean of means is the grand mean.

Note that when the means are calculated from subsets of different sizes, they should be weighted to obtain the grand mean. Thus if the mean of a group of five observations is 8, and the mean of another group, consisting of fifteen observations is 12, the grand mean is

$$\frac{(5\times8)+(15\times12)}{20}=\frac{220}{20}$$

$$=11$$

or, more simply

$$\frac{(1\times8)+(3\times12}{4}=\frac{44}{4}$$

$$=11$$

Graph. The presentation of statistical data in geometric form. This use of visual image facilitates the comprehension of the pattern of distribution of a variable (a characteristic of the population studied) or the relationship of two or more variables. Differences in quantities are represented by the area, shape, color, or other differentiating characteristics of the design.

Graphical Estimator. A constant which was chosen by trial and error from a geometrical representation generally for securing a linear plot of observations on a specially ruled paper.

Grid. Two sets of parallel equally spaced lines drawn such that the sets of lines intersect each other vertically. A grid may consist of horizontal and vertical lines or of diagonally drawn intersecting lines. Grids may be used in area sampling to select particular subjects of an area for specific examination.

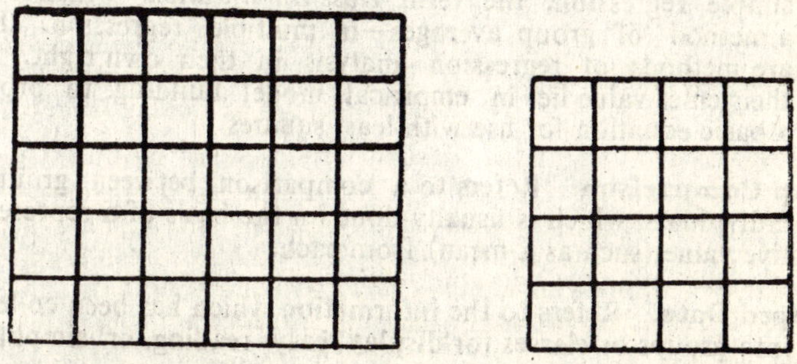

Examples of grids

Grid Sampling. Refers to a form of cluster sampling, the clusters being individual areas of a grid and henee having groups of basic cells which are arranged in some standard geometrical pattern. The term 'configurational sampling' has been also used in the same sense.

Gross Errors. The total errors involved in the measurement or observation process. The examination of gross errors is particularly important in methodological studies where we are interested in the process of measurement itself. Whereas estimates of population means or totals are affected only by net errors—that part of the total error which does not cancel our over all the observations—measures of association are affected by gross errors. In addition, any attempt to eliminate errors must be based on a knowledge of the gross errors involved and the circumstances under which they arise.

Grossing Up. A process used for estimating the total for a universe from a sample whose proportion of the universe is known precisely or approximately. The multiplier used for the purpose is known as the *grossing-up factor*, which, for a 5% sample, would be 20.

Group. Refers to a set of elements, individuals or observations all of which possess one, or more, characteristics in common. The word is also found occasionally in statistics in its mathematical sense.

Group Averages. A method of simple-regression analysis in which the sets of observations are arrange in two or more groups of equal size in accordance with the magnitude of the independent variable averaged, and either plotted in a graph usually with the individual items, or analysed by finite differences. The term *group averages* is used in reference to simple regression, the term *cross-classification*—which is also a method of group averages—in multiple regression. Both are methods of regression analysis in their own right. But their chief value lies in empirical model building to provide a basic equation for use with least squares.

Group Comparison. Refers to a comparison between groups of individuals, which is usually done on the basis of a representative value (such as a mean) from each.

Grouped Date. Refers to the information which has been collected into groups or classes for display, easy reading, or simplifying calculation.

Example: the heights of 100 waves had been recorded one day at a coastal research station. They were as follows (in m) :

.701	.192	1.002	.824	.702	.201	.889	.546	1.082	1.103
1.013	1.271	.212	.842	.936	.836	.402	.963	.614	.297
.303	.757	.768	.324	.959	.863	1.112	1.173	.812	1.054
.867	.942	.714	.723	1.521	.912	1.223	.625	1.145	.161
.746	.182	.646	1.216	.496	.934	.433	1.046	.419	.688
.736	.296	1.289	1.024	1.362	.946	1.292	.679	.836	.781
.573	.846	.584	.637	.991	.450	.476	.812	.369	.849
.836	.658	.742	1.039	.691	.853	1.082	.529	.962	1.330
.739	1.154	.492	.929	1.234	.824	.770	1.341	1.356	.538
1.256	.516	.669	.550	1.190	.998	.199	.792	.479	.901

Looking at these figures as they stand gives very little immediate information to the reader. When, however, they have been grouped in classes they are much easier to interpret.

Height (*m*)	0−<.2	.2⩽<.4	.4⩽<.6	.6⩽<.8
Frequency	4	7	15	22
Height (*m*)	.8⩽<1.0	1.0⩽<1.2	1.2⩽<1.4	1.4⩽<1.6
Frequency	22	14	11	1

The mean *x* and standard deviation of ground data have been estimated by assuming all the members of each group to be at the mid-point of the group.

Height (*m*)	*Mid-point* x	*Frequency* f	xf	$(x-\overline{x})$	$(x-\overline{x})^2$	$(x-\overline{x})^2f$
.0−.2	.1	4	.4	−.7	.49	1.96
.2−.4	.3	7	2.1	−.5	.25	1.75
.4−.6	.5	15	7.5	−.3	.09	1.35
.6−.8	.7	22	15.4	−.1	.01	0.22
.8−1.0	.9	26	23.4	.1	.01	0.26
1.0−1.2	1.1	14	15.4	.3	.09	1.26
1.2−1.4	1.3	11	14.3	.5	.25	2.75
1.4−1.6.	1.5	1	1.5	.7	.49	.49
Total		100	80			10.04

$$\text{Mean } \overline{x} = \frac{\Sigma xf}{\Sigma f} = \frac{80}{100} = .8 \text{ (estimated)}$$

$$\text{Variance } s^2 = \frac{\Sigma(x-\overline{x})^2 f}{\Sigma f} = \frac{10.04}{100} = .1004$$

$$\text{SD} = \sqrt{\text{Variance}} = .32$$

If the distribution is having a central mode and tails off to either side, as in this case, this procedure will involve errors.

For groups below the mode, more of the members of a group would be above the mid-point than below it (Fig.). For groups above the mode, the opposite would be true.

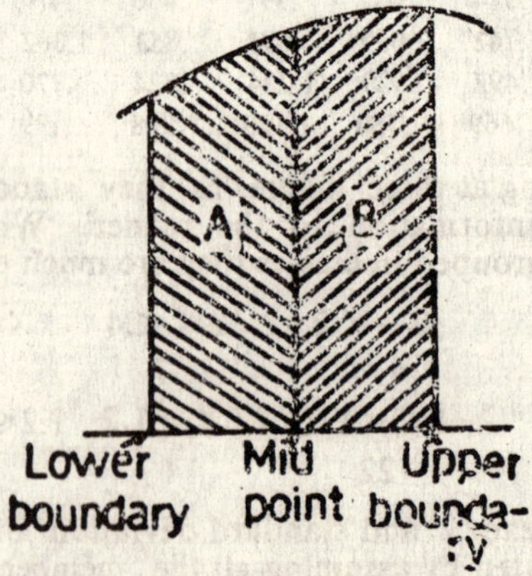

Grouped data. The population B, above the mid-point of this class, is greater than that of A, below the mid-point. This is the case for any class below the mode.

For estimation of the mean, the errors above and below would approximately cancel each other out, so the method described has been reasonably accurate. For standard deviation, however, the errors will add up so that the estimated value would be slightly too large (some of the population is nearer the middle than estimated). This could be allowed for by applying *Sheppard's correction* :

$$\textit{Corrected variance} = \text{calculated variance} - \frac{c^2}{12}$$

where c has been the class interval

For the figures in the example of wave heights :

$$\textit{Corrected variance} = .1004 - \frac{(.2)^2}{12} = .0971$$

$$\textit{Corrected standard deviation} = .312m$$

Seppard's correction has been only of value if the distribution has been approximately normal. Thus it must be continuous (or discreate with no gaps in the data), and must tail off to zero either side of the mode. The class interval should be less than the standard deviation.

Group Factor. See common factor.

Grouping Lattice. A lattice or mesh of equal variate intervals which, superimposed on a variate scale, defines the intervals in which the frequencies have been grouped.

Group Internal-Control Matrix. A matrix or table designed to facilitate the control of the inter-divisional transactions of a company or the intercompany transactions of a group. Basically, it consists of an inter-output matrix internalto the company or group.

Grouped Poission Distribution. In the pure birth stochastic process if it is assumed that the time between two events is having the gamma distribution then the population size at a given time is having the grouped Poission distribution with parameters k and λ with probability function

$$\Pr(x)=\sum_{j=1}^{k} e^{-\lambda}\frac{\lambda^{xk+j-1}}{(xk+j-1)!} \quad \text{where } x=0, 1, 2, \ldots, k$$

has been a positive integar and λ has been positive.

Growth Curve. In general, this curve refers to an expression which gives the size of a population y as a function of a time-variable t, and hence describing the course of its growth. The expression may also be used for denoting the growth of an individual.

Gurland's Generalisation of Neyman's Distribution. Refers to the mixture of Neyman's type A distribution having two parameters λ_1 and $\lambda_2 p$, where p is having the beta distribution with parameters α and β, is Gurland's generalisation of Neyman's distribution with parameters λ_1, λ_2, α and β where all four parameters are positive. The Beall-Rescias generalisation has been a special case of this Gurland generalisation.

H

Half-Cauchy Distribution. Refers to a special case of the folded Cauchy distribution which, in standard form is having the density function

$$\frac{2}{\pi(1+x^2)} \quad (0<x).$$

This distribution has been analogous with the half Normal distribution.

Half Normal Distribution. Refers to a special case of the folded Normal distribution. If the distribution of x has been $N(0, \sigma)$ then the distribution of $z = |x|$ is 'folded' in half and has the density function.

$$f(z) = \frac{1}{\sigma}\sqrt{\frac{2}{\pi}}\, e^{-z^2/2\sigma^2}.$$

Half Normal Plots. Refers to a graphical method for interpreting the contrasts in a two level factorial experiment. By ranking these contrasts they could be shown as a cumulative distribution function on a modified form of arithmetic probability paper. The modification consists of suppressing the lower half of the probability scale and replacing each value on the upper half by $P' = 2P - 100$. The ordinary plotting convention of $P^1 = (i - \frac{1}{2})/n$ has been then used.

Half-Normal Probability Paper. Normal probability paper with negative abscissas omitted.

Half-Replicate Design. Refers to an experimental design based upon the principle of fractional replication, which uses only one half of the complete number of treatment combination in a basic design.

Half-Width. This expression has been sometimes used in relation to central confidence intervals for denoting the upper or lower half of an interval. In fixed interval prediction, such as may be used in a control chart, the half-width refers to the distance on the scale of the variate between, say, the process average and the upper or lower control limit.

An objectionable use of the term sometimes takes place in elementary texts. If a frequency function has been unimodal with a frequency f_0 at the mode, and if there have been two points x_1 and x_2 where

$$f(x_2) = f(x_2) = \tfrac{1}{2} f(x_0)$$

the distance $\frac{1}{2}(x_2 - x_1)$ has been sometimes called the half-width.

Haphazard Sampling. Selecting the sample for observation without taking particular care about the representativeness of the elements selected. Samples of volunteer subjects are haphazard samples. It is not appropriate to make references about the whole population on the basis of such samples. The term should not be confused with random sampling, in which every element in the population is having a known non-zero probability of selection.

Hardy Summation Method. Refers to a method of determining the moments of a frequency function defined at equidistant points, or grouped in equal intervals, by repeated summation of the frequencies. It was given by Sir George Hardy in 1903. The summations provide factorial moments from which ordinary moments may easily get derived.

Harmonic Analysis. Refers to the analysis of a series of values into constitutent periodic terms.

Harmonic Distribution. A term to be avoided.

Harmonic Mean. A special kind of mean which is determined by calculating the mean of the reciprocals of the values concerned and then calculating the reciprocal of this quantity.

The harmonic mean m of the n numbers $x_1, x_2, \ldots x_n$ is given by :

$$\frac{1}{m}=\frac{1}{n}\left\{\frac{1}{x_1}+\frac{1}{x_2}+\ldots+\frac{1}{x_n}\right\}$$

Thus, the harmonic mean of 2, 4 and 5 may be calculated :

$$\frac{1}{m}=\frac{1}{3}\left(\frac{1}{2}+\frac{1}{4}+\frac{1}{5}\right) \text{ giving } m=\frac{60}{19}=3\frac{3}{19}$$

The harmonic mean may be thus also given by :

$$\frac{n}{\sum_{i=1}^{n}\frac{1}{x_i}}$$

Example : a man travels 1 km at 5 km/h, 1km at 10km/h and 1km at 20 km/h. What is his average speed ?

$$\textit{Average speed}=\frac{\text{Distance travelled}}{\text{Time taken}}$$

$$=\frac{3\text{km}}{\left(\frac{1}{5}+\frac{1}{10}+\frac{1}{20}\right)\text{h}}=8\frac{4}{7}\text{ km/h}$$

Thus, the average speed would be the harmonic mean of the three speeds.

A problem which may be solved by using the harmonic mean is given in the example.

Harmonic Process. A stationary random process which satisfies the relationship :

$$x_t+b_1x_{t-1}=b_2x_{t-2}+\ldots+b_hx_{t-h}=0$$

has been a harmonic process. It is deterministic, the general solution being a sum of harmonic terms which may get damped or explosive.

Harris Walk. Refers to a random walk on the set of non-negative integers, for which the matrix of transition probabilities is consisting of zeros except for the elements.

$$P_{0,1}=1$$

$$P_{j,\ j+1}=a_j \qquad 0<a_j<1,\ 1\leqslant j<\infty$$

$$P_{j,\ j-1}=1-a_j$$

Hazard. In general, refers to a word which implies the existence of chance of risk (from Arabic 'al zhar', meaning a die). Specialised usage takes place principally in connection with lief analysis of physical systems or components. The hazard has been the probability that an article, functioning at time t_0 will fail in the interval $(t,\ t+\delta t)$ and this is given by $h(x)=f(x)/[1-F(x)]$. If this instantaneous (death or failure) rate has been calculated at a number of survival times, the resulting hazard rates may be considered as a hazard function.

Helmert Distribution. Refers to the distribution of the sample standard deviation or, equivalently, of the sample variance in samples from a normal population. It may be written as follows :

$$dF=\frac{n^{\frac{1}{2}(n-1)}}{\lambda\{\frac{1}{2}(n-1)\}2^{\frac{1}{2}(n-3)}}\left(\frac{s}{\sigma}\right)^{n-2}e^{-ns^2/2\sigma^2}\frac{ds}{\sigma},$$

$$0\leqslant s\leqslant\infty$$

where σ^2 denotes the parent variance and s^2 the sample variance. If ns^2/σ^2 has been put equal to χ^2 the distribution becomes that of χ^2 with $n-1$ degrees of freedom. It has been a form of the type III distribution.

Hermite Distribution. The probability generation function of a generalised Poisson distribution may be put as $g\ (z)=$

$\exp\{a_1(z-1)+a_2(z^2-1)+\ldots\}$ where $\sum_{i=3}^{\infty} a_1=\lambda$, the Poisson parameter. This generalised distribution tends to become the Poisson distribution having parameter a_1 if a_2, a_3... become negligible compared to a_1 in some limiting process. If a_2 could not be neglected as compared to a_1, then limiting distribution would be the Hermite distribution with parameters a_1 and a_2.

Heteroscedasticity. Scedasticity is a term, which is not much used. It denotes dispersion, particularly in terms of variance. The term heteroscedasticity finds use in regression analysis to describe a situation where the variance of the error or disturbance term differs for different values of the independent variable. In general, the distribution of a variable has been said to be heteroscedastic if its variance differs for different fixed values of another variable. If the variance has been the same for all fixed values of the other variable, the distribution has been said to be homoscedastic. For example, the plot below shows a case in which the disturbance term is having a variance depending on thc valne of x, *i.e.* $y=\alpha+\beta x+\epsilon x$.

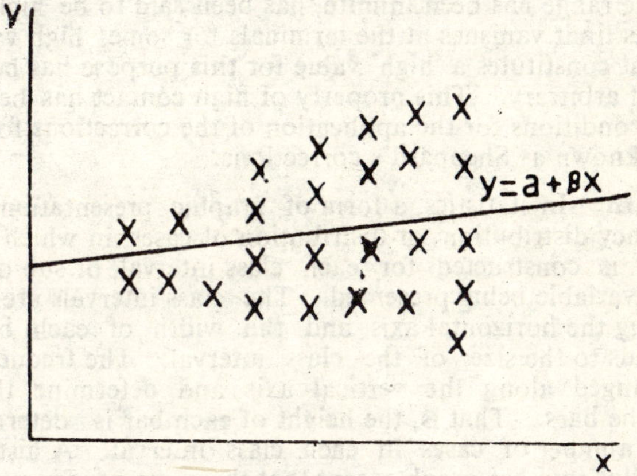

Heterotypic. Refers to a term which is used in relation to Pearson distributions. For certain values of the moment ratics β_1 and β_2 the differential equation defining the family of distributions is having infinite moments of orders 8 or more. In this region the standard error of the sample estimate of β_2 would be infinite and hence the fitting of the distributions by the method of moments would get inappropriate. Distributions of the Pearson family with such values of β_1 and β_2 had been termed as hetrotypic ; but, as is now realised, certain of the distributions may get valuable for many purposes.

Hierachy. If, in a matrix of intercorrelations of a set of variates, the rows and columns could be so arranged to provide the highest correlations in the upper left-hand corner and the lowest correlations in the lower right-hand corner and when this is performed there. has been a constant proportional relationship between adjacent columns, except for diagonal terms, the table has been called a hierarchy (Spearman, 1904) and the intercorrelations are said to be an hierarchical order. Thus' for two rows denoted by a and b and two columns by c and d, the correlations are obeying the so-called tetrad relations $r_{ac}r_{bd}=r_{ad}r_{bc}$. Under certain conditions the fact that the correlation matrix has been hierarchical has been a necessary and sufficient condition that the variation can be accounted for by a single factor common to the variates.

The quantities $r_{ac}r_{bd}-r_{ad}r_{bc}$ are termed as tetrade differences.

High Contact. In relation to a frequency function $f(x)$, the order of contact of the function having the variate axis, or at infinity if the range has been infinite, has been said to be high if $x^r f(x)$ or its limit vanishes at the terminals for some high value of r. What constitutes a 'high' value for this purpose has been somewhat arbitrary. This property of high contact has been one of the conditions for the application of the corrections for grouping known as Sheppard's corrections.

Histogram. In statistics, a form of graphic presentation of a frequency distribution, or distribution of cases, in which a vertical bar, is constructed for each class interval, or sub-division of the variable being presented. The class intervals are arranged along the horizontal axis and the width of each bar corresponds to the size of the class interval. The frequencies are arranged along the vertical axis and determine the height of the bars. That is, the height of each bar is determined by the number of cases in each class interval. A histogram is similar to a bar graph except that there are no spaces between the bars.

Historigram. The graph of a time series with the series measured on the verticle scale, and time measured on the horizontal scale. It is common practice to connect adjacent dots by straight lines and not to attempt to fit a smoothed curve through the dots themselves, though a general curve can be fitted by regression analysis or by moving averages. The distance between scale points on the time scale represents the lapse of time rather than periods, so that lines drawn from the calibration marks of the time scale should represent either periods or points of time for rate-of-flow figures and point-of time figures, respectively ; the plotted dots thus fall on the lines and not between them. Owing to possible confussion with the word 'histogram' the term is not to be recommended.

Homogeneous/Homogeneity. The term has been used in statistics in relation to samples from different populations which may or may not be identical. If the populations have been identical, they are said to be homogeneous, and thus the samples drawn from these populations are also said to be homogenous. It has been more common to restrict the examination to particular characteristics of the populations, *e.g.*, the means or variances. There a number of populations with identical means are homogeneous in their means.

Homogeneous Process. A stochastic process is considered to be homogeneous in space if the transition probability between any two state values at two given times is depending only on the difference between those state values.

The process has been homogeneous in time if the transition probability between two given state values at any two times depends only on the difference between those times.

Homograde. A term used to denote a qualitative variate *i.e.*, an attribute. The word has been not in common current use among English writers.

Homoscedastic. See homo scedasticity.

Homoscedasticity. Scedasticity is a term, not widely used, which denotes dispersion, particularly in terms of variance. The term homoscedasticity is used in regression analysis of describe a situation where the variance of the error term or disturbance term is the same for all values of the independent variable. In general, the distribution of a variable is said to be homoscedastic if its variance is the same for all fixed values of another variable. If the variance differs for different values of the other variables, the distribution is said to be heteroscedastic.

Hunt-Stein Theorem. Refers to a theorem concerning critical regions which are associated with a particular significance test. For example, in the set of critical regions of size *a* which have been also invariant, there may be one, say, W_0, which has been uniformly most powerful and most stringent.

Hybrid Series. A statistical series consisting of a mixtue of time series figures and cross-section figures, sometimes called more precisely a *temporal cross-section series.* It is adopted for analytical purposes where : (1) enough data of one kind or the other are not available ; (2) cross-section data are limited in number of observations and have to be supplemented by time data ; (3) in a regression analysis, cross-section data are preferred but are available for some variables only, and time series data have to be used for other variables. In the latter case, two separate analyses are carried out, and the two resulting regression equations merged into one.

Hyperbolic Secant Distribution. Refers to a frequency distribution which gets related to the logistic distribution, with probability density function

$$\Pr_y(y) = \pi^{-1} \operatorname{sech} y$$

and cumulative distribution function

$$F_y(y) = \tfrac{1}{2} + \pi^{-1} \tanh^{-1} (\sinh y).$$

Hyperexponential Distribution. A term which was proposed by Morse (1958) to a mixture of exponential distributions in the general form :

$$\gamma e^{-2\gamma bt} + (1-\gamma)\, e^{-2(1-\gamma)bt}.$$

Hypergeometric Distribution. The probability distribution $p(x)$ may be given as follows :

$$p(x) = \frac{{}^{a}C_{x}\,{}^{b}C_{n-x}}{{}^{a+b}C_{n}}$$

whree $0 \leqslant x \leqslant n$; $n \leqslant a+b$: a and b are positive integers, and ${}^{a}C_{x}$, etc., are binomial coefficients.

For this distribution,

$$Mean = \frac{na}{a+b}$$

$$Variance = \frac{nab\,(a+b-n)}{(a+b)^2(a+b-1)}$$

The distribution has been generated when a sample of size n has been taken (without replacement) from a parent population containing a items of one particular type, and b of others. The probability that the sample is having exactly x of the particular type is given by the relevant term in the hypergeometric distribution.

Hypergeometric Waiting Time Distribution. Refers to an alternative name for the inverse hypergeometric distribution.

Hyper-Poisson Distribution. This discrete distribution is having a complicated probability function in two parameters, λ and θ.

If $\lambda=1$ this distribution becomes the Poisson distribution with parameter θ. If $\lambda<1$ it has been sometimes referred to as a sub-Poisson distribution and if $\lambda>1$ as a super-Poisson distribution.

where

$$c_p=\left[\int_0^{\pi}(\sin^{p-1}t)\exp(k\cos t)\,dt\right]^{-1}$$

For the case $p=3$, the distribution has been sometimes termed as the spherical Normal distribution.

Hypothesis. In scientific investigation, a tentative statement asserting a relationship between certain facts. The statement is intended to be tested empirically and either verified or rejected. A scientific hypothesis is derived from a theoretical system and the results of past research. The clear statement of a theoretically significant hypothesis related to previous research is crucial to a well-designed scientific study. See experiment, theory.

Hypothesis, Statistical. A hypothesis which is concerning the parameters or form of the probability distribution for designated population or populations, or, more generally, of a probabilistic mechanism which has been supposed to generate the observation.

Hypothetical Population. Refers to a statistical population which is having no real existence but is imagined to be generated by repetitions of events of a certain type *e.g.*, the binomial distribution as generated by the throws of a die, or crop-yields on a set of plots imagined as all the possible ways in which a set of yields could occur under the conditions of an experiment.

I

'Ideal' Index Number. In 1922 Irving Fisher put forward certain criteria which have to be obeyed by 'good' index numbers. Of the large collection of formulae investigated only a few have been found to obey his tests. One of these was termed the 'ideal' index. It may be but as follows :

$$\left\{\frac{\Sigma p_n q_o}{\Sigma p_o q_o}\times\frac{\Sigma p_n q_n}{\Sigma p_o q_n}\right\}^{1/2}$$

where p_o, q_o denotes prices and quantities in the base period and p_n, q_n those of the period for which the index is being calculated. The ideal index has been the geometric mean of the Laspeyres' and Paasche index numbers.

Identical Errors. A regression model of type $y=f(x_1,..., x_p)+\epsilon$ where ϵ possesses a unique frequency distribution, is considered to have identical errors. 'Identity' refers not to individual errors but to their frequency distribution.

Illusory Association. An association between attributes may be fometimes said to be illusory if it may be statistically significant without necessarily involving any direct causal connection between them.

Imbedded Process. A stochastic process in continuous time has been observed only at the time points where a change of state takes place. These points of discontinuity can be considered to be forming a new discrete time variable.

Imperfect Competition. A degree of competition which affords the seller an appreciable amount of control over his selling prices.

Implied Elasticity of Demand. A null hypothesis, to the effect that the current price of a brand of a firm's products is rational based on the postulation that the implied price elasticity of demand is the true one. The formula of the rational price or *optimum price*, as it is often called, as

$$p_m = a\,\frac{e}{e-1}$$

where p_m is the rational price, a the marginal cost, and e the elasticity of demand.

Implicit Strata. Refers to the n groups, zones or sub-classes into which, in systematic sampling, the population has been divided by the sampling interval $k=N/n$, where N denotes the population size and n the sample size.

Importance Sampling. This method of sampling was used in Monte Carlo analysis and was used to concentrate the distribution of sample points in those parts of the variate interval which have been of greatest interest. It has been a form of stratified sampling having variable sampling fraction.

Impossible. Incapable of occurring of happening. If an event is impossible, the probability of its occurrence is zero.

Improper Distribution. Refers to a probability distribution having a distribution function F(x) which F($-\infty$)$>$0 or F($+\infty$)$<$1. Such distribution are having adherent probabilities at $-\infty$ or $+\infty$ respectively.

Incidence Rate. The number of spells beginning during the observation period retated to the average number of persons under observation.

Incomplete Beta Function. This function may be defined as follows ;

$$B_t\ (s, r = \int_0^t y^{s-1}\ (1-y)_r^{-1} dy.$$

$$s'r>0\ ;\ 0 \leqslant t \leqslant 1.$$

Incomplete Census. See census.

Incomplete Coverage. A survey or census in which a considerable number of the elements in the population are not included may be called incomplete.

Incomplete Gamma Function. A function which is defined as follows :

$$\lambda_t\ (\lambda) = \int_0^t e^{-x} x^{\lambda-1} dx\ \lambda>0\ ;\ 0 \leqslant t \leqslant \infty.$$

It has been the distribution function of the gamma distribution multiplied by $\lambda(\lambda)$.

Incomplete Moment. The ordinary moment of a distribution about an arbitrary origin a may be given by

$$\mu_r' = \int_{-\infty}^{\infty} (x-a)^r\ dF(x).$$

Increment, 1. Refers to a small change in the value of a variable. 2. Refers to change the value of a variable systematically by small amounts.

Incremental Analysis. Refers to the process of determining an optimum value by building up a statistical schedule.

Independence. This term in the calculs of probabilities, may be generally defined by reference to the principle of compound probabilities. Two events are considered to be independent if the probability of one has been the same whether the other has been given or not, *i.e.*, $Pr(A) = Pr(A \mid B)$ and $Pr(B) = Pr(B \mid A)$.

From this it is evident that the probability of the compound event $Pr(AB) = Pr(A)\ Pr(B)$ if the events are independent. As a matter of axiomatisation it is desirable use relations of the type $Pr(AB)=Pr(A)\ Pr(B)$ as definitions to avoid difficulties arising when $Pr(A)$ or $Pr(B)$ is zero.

In statistics two variates x_1 and x_2 have been independent if their distribution functions get related by

$$F(x_1, x_2)=F(x_1, \infty)\ F(\infty, x_2)$$

or equivalently if their frequency functions, should they exist, are related by $f(x_1, x_2,)=f_1(x_1)\ f_2(x)._2$ Generally n variates $x_1, x_2. . ., x_n$ have been independent if

$$F(x_1, x_2, . . ., x_n)=F(x_1, \infty, \infty, . . ., \infty)$$
$$F(\infty, x_2, \infty . . ., \infty). . .F(\infty, \infty, \infty, . . ., x_n).$$

It has not been enough that they should be independent pair and pair. The word has been also applied in the ordinary mathematical sense to describe the independence of two or more variables.

Independence Frequency. In a contingency table, this term refers to the frequency which would be found in a particular cell if the attributes defining it had been independent; *e.g.*, if the rows of the table are having total frequencies $A_1, A_2. . ., A_r$ and the s columns are having frequencies $B_1, B_2, . . . B_s$, and

$$\Sigma A=\Sigma B=N,$$

the independence frequency in the ith row and jth column would be $A_i B_j/N$.

Independent Events. Refer to the events which have no influence on each other. If a die and a coin have been thrown together, the events 'The die shows 6' and 'The coin comes heads' have been independent.

This is expressed mathematically by saying that event A has been independent of event B if

$$p(A \mid B)=p(A \mid B')=p(A)$$

In the same way, event B is independent of event A if

$$p(B \mid A)=p(B \mid A')=p(B)$$

In fact, if A is independent of B, then B must be independent of A and

$$p(A \cap B)=p(A)\ p(B)$$

Independent Random Variables. Refers to the random variables, *e.g.*, x and y, such that a knowledge of the value of x does not inference the probability distribution of y, and vice versa. Thus there is no relationship between the values of independent random variables, and their covariance is zero.

Independent Trials. The successive trials of an event have been considered to be independent if the probability of outcome of any trial has been independent of the outcome of the others. The expression has been usually confined to cases where the probability has been the same for all trials.

Independent Variable. This term is used in regression analysis to mean one or a number of predictor or explanatory variables (or variates) from which a dependent variable is said to have resulted. It is a convention to use the letter Y to indicate the dependent variable and $X_1, X_2, X_3, \ldots X_n$ for each of the independedt variables. Thus a relationship between independent variables may be defined

$$Y=\beta_0+\beta_1X_1+\beta_2X_2+\beta_3X_3 \ldots +\beta_nX_n+\epsilon$$

Index. Any measurable or observable phenomenon (or phenomena) that is used to indicate the presence of another phenomenon that cannot be measured directly or conveniently. Thus one may attempt to determine the social status of individuals in a social system by determining their occupations. Occupation, then, is used as an index of social status.

Index Numbers. A statistical series denoting relative values, usually where there is no absolute measure, and almost invariably standardized to one item of the series given as 100. The time or place given as equalling 100 is called the base.

In *Laspeyre's Index* the weighting has been taken from the base year, in *Paasche's Index* from the current year. This implies that Laspeyre's Index tends to emphasize items which are out of date and to ignore changes in taste or life-style over the years. *Paasche's index*, on the other hand, runs into difficulties with new items which are significant in the current year but have no base value because they did not exist or were not used before. Imagine, for example, a cost of living index, set up with base year 1800 to compare the cost of living then with now. If Laspeyre's index has been used, items like candles and horse feed will be quite heavily weighted, whereas electricity and petrol would not feature at all. If Paasche's index has been used, the reverse has been the case.

A compromise between the two has been the *Typical Year Index*, where the weightings have been taken for some typical time between the base year and the present.

In *Fisher's Ideal Index*, the value has been taken as the geometric mean of Laspeyre's Index and Paasche's Index. The *Marshall Edgeworth Bowley Index* takes the weighting for any item to be the average of that for the base year and that for the current year.

Formulae for the various indexes are as follows, where p_o=base year price, p_n=current year price, q_o=base year weighting, q_n=current year weighting, and q_b=typical year weighting :

$$\textit{Laspeyre's Index}=\frac{\Sigma p_n q_o}{\Sigma p_o q_o}$$

$$\textit{Paasche's Index}=\frac{\Sigma p_n q_n}{\Sigma p_o q_n}$$

$$\textit{Typical Year Index}=\frac{\Sigma p_n q_t}{\Sigma p_o q_t}$$

$$\textit{Fisher's Ideal Index}=\sqrt{\left\{\frac{\Sigma p_n q_o}{\Sigma p_o q_o}\right\}\left\{\frac{\Sigma p_n q_n}{\Sigma p_o q_n}\right\}}$$

Marshall Edgeworth

$$\textit{Bowley Index}=\frac{\Sigma p_n\ (q_o+q_n)}{\Sigma p_o\ (q_o+q_n)}$$

Indirect Effect. Refers to that part of the effect of one variable on another in a path model which operates through its effect on an intermediate variable in the model.

Indirect Sampling. Refers to sampling from documents, or some record of the characteristics of a population, rather than the recording of information obtained at first hand from units of the population themselves.

Individual Response Error. The individual response error is the difference between the individual true value and the information recorded for the individual.

Inefficient Statistic. Refers to a statistic having less than the maximum possible efficiency in the sense of having the minimum sampling variance.

Inferential Statistics. Refers to that branch of statistics which involves drawing conclusions from data by hypothesis testing.

Infinite Divisibility. A distribution F has been infinitely divisible if and only if for each n it could be represented as the distribution of the convolution (sum)

$$S_n = X_{1,n} + \ldots + X_{n,n}$$

of n independent random variables having a common distribution F_n.

Infinite Population. An infinite population may be defined as one which either has the infinite property through some limiting process or, in a sampling context, can be made to possess that property by some strategy of sampling.

Information. A term which is having a particular meaning in the theory of estimation.

The amount of information about a parameter ϕ from a randomly drawn sample of n independent observations, where the population from which the sample is drawn has a frequency function $f(x, \phi)$ can be defined as

$$n\,\epsilon\left(\frac{\delta \log f}{\delta\phi}\right)^2 \equiv n\int_{-\infty}^{\infty}\left(\frac{\log f(x.\ \phi)}{\delta\phi}\right)^2 f(x, \phi)\,dx$$

In general it may be said that if the variance of unbiased estimators of ϕ is large the information is small, and that if the variance is small the information is large.

Information Theory. A mathematical theory of communication which is concerned with problems relating to the accuracy, efficiency, and effectiveness of communication. It deals with questions such as the extent to which the symbols used in communication convey the intended meaning, the effect of extraneous disturbances (referred to as "noise") in introducing errors in communication, the extent to which there is "redundancy" in communication—that is, portions of a message not needed to convey the intended meaning.

Input-output Analysis. A technique in which the output of each industry in an economy is analysed according to the use made of it. It can be used to calculate the level of output in different industries necessary to sustain a given pattern of final demand.

Input/Output Process. Refers to a broad class of stochastic processes, the stale $\xi(t)$ at time t being the resultant of two random variables x, y;

$$\xi(t) = x(t) - y(t),$$

x and y being called respectively the input and the output. Examples have been queueing processes, provisioning and epidemic processes.

Inspection Lot. A lot presented for inspection, which could be carried out on each member of the lot or on a sample of members only.

Instantaneous Death Rate. Refers to a version of the age specific death rate where the 'next unit time period' has been reckoned to be short.

Instantaneous State. Refers to a state of a Markov chain in continuous time for which the expected sojourn time has been zero.

Integer Programming. Refers to a form of linear programming where the solutions get derived in an integer form, *e.g.*, in the scheduling of ships or men.

Integrated Data. Refers to a class of statistical data in which the values for short unit intervals could be added together to provide a series of values relating to longer intervals.

Intelligence Quotient, IQ. Refers to a measure of the intelligence of an individual. Intelligence quotient may be given by :

$$IQ=\frac{\text{Mental age}}{\text{Chronological age}}\times 100$$

Intelligence quotient is having mean value 100, standard deviation 15.

Intensity. In the harmonic analysis of time series, this term refers to a measure which gives an estimate of the amplitudes of the constituent harmonics. If the series has been u_1, $u_2, \ldots, u_n$ the intensity for a period μ may be defined as A^2+B^2 where

$$A=\frac{2}{n}\sum_{j=1}^{n} u_j \cos\frac{2\pi j}{\mu},\ B=\frac{2}{n}\sum_{j=1}^{n} u_j \sin\frac{2\pi j}{\mu}.$$

Intensive Sampling. Like extensive sampling this expression may imply two different things : either (*a*) sampling in a particular area with a dense scatter of sampling points or (*b*) sampling wherein information on a restricted range of topics has been sought by probing on them very deeply with an intricate schedule of questions.

Interaction. When the effect of a change in one variable differs for different levels of another variable there is said to be an interaction between them.

Intercorrelation. A term which is used to denote the correlation of a number of variates among themselves, as distinct from the correlations between them and an 'outside' or dependent variate.

Interdecile Range. The range of observations of a variate between the first and ninth deciles.

Intermediate Products. Goods at an intermediate state of production—neither raw material as purchased by a manufacturing firm, nor the finished article ready for sale to customers ; an important concept in the censuses of production.

Intergraduated Values. Values which have been subject either to cograduation or to contragraduation.

Internal Rate of Return (IRR). That rate of discount which converts the total yield—gross of depreciation—of a capital project, to a present value equal to the capital outlay on the project.

Interpenetrating Samples (subsamples). With interpenetrating sampling, instead of selecting a single sample from the population we select a number of subsamples using the same sample design for each. Each subsample must itself be a self-contained sample of the population. Interpenetration may be used with any sample design and at any stage of the selection.

Interpolation. The process of calculating an unknown value of a variable that is between two or more known values in the series, When the known values are plotted on a graph, the process of interpolation assumes that the unknown value lies on the line of the graph.

Interpretive Statistics. Refers to that branch of statistics which is concerned with drawing conclusions from data.

Interquartile Range. Refers to the variate distance between the upper and lower quartiles. This range is having one half of the total frequency and provides a simple measure of dispersion which finds use in descriptive statitics.

Interrupted Distributions. A name assigned to distributions which are arising in circumstances in which some values of a random variable cannot be observed, *e.g.*, radio-active particles arriving in the 'dead' time during which the counting instrument, having just recorded the arrival of a particle, has been unable to record further arriving particles. The distribution of particles recorded therefore is different from the distribution of actual arrivals.

Intersection. Refers to the elements that have been common to two (or more) sets (Fig.). The intersection of sets A and B has been written $A \cap B$. For example, if A $\{s, t, a, r\}$ and $B = \{p, l, a, n, e, t\}$, then $A \cap B = \{a, t\}$.

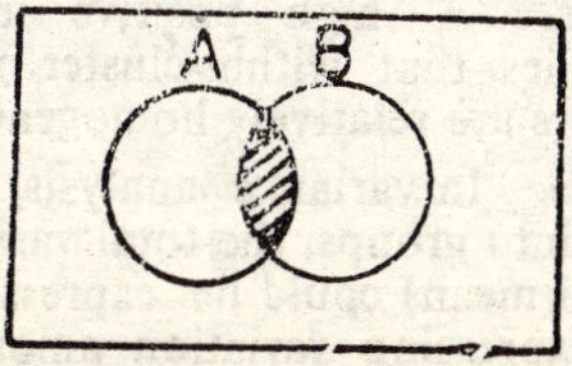

$A \cap B$ is the shaded area.

Interval Distribution. Refers to a probability distributin which is formed by the intervals, or gaps, in time or distance between events or occurrences.

Interval Estimation. Refers to the estimation of a population parameter by specifying a range of values bounded by an upper and a lower limit, within which the true value has been asserted to lie, as distinct from point estimation which assigns a single value to the true value of the parameter. The unknown value of the population parameter has been presumed to lie within the specified interval either on a stated proportion of occasions, under conditions of repeated sampling, or in some other probabilistics sense.

Interview. A structured social interaction between two people—an interviewer whose task it is to obtain information, and a respondent whose function it is to provide this information.

Interviewer Effect. A term sometimes used to denote the totality of errors in survey responses due to the interviewer. It includes therefore interviewer bias and interviewer variance.

Intraclass. Within a class or group. We distinguish between intraclass and interclass. In the analysis of variance, for example, the total sum of squares can be partitioned into the interclass (as between groups) sum of squares and the interclass (or within group) sum of squares.

Intraclass Correlation. Refers to a measure of correlation within the members of certain natural groups or 'families'.

Intraclass Correlation Coefficient. A measure of the internal homogeneity of a class, category, group of family of elements.

Intracluster Correlation Coefficients. A synonym for intraclass correlation coefficient. The term 'intracluster' is used mainly in connection with survey sampling where the 'cluster' are the groups of elements selected in cluster sampling. The coefficient is usually positive for human populations, since people living in the same neighbourhood tend to be more similar to one another than they are to people living in other neighbourhoods. A high positive intracluster correlation coefficient implies that within-cluster variability is relatively low—the clusters are relatively homogeneous.

Intraclass Variance. In variance analysis, where the data have been classified into groups, the total variation (sum of squares about the grand mean) could be expressed as the sum of two components, expressing variation among for means of groups and variation within groups. The latter has been the sum of squares about group means pooled for the various groups, and an estimate of variance within groups based on it is termed as within group or intraclass variance.

Intrinsic Accuracy. The intrinsic accuracy of a distribution with frequency function $f(x, \theta)$ may be defined as follows :

$$I=\int_{-\infty}^{\infty}\left(\frac{\partial \log f}{\partial \theta}\right)^2 f(x)\,dx.$$

Strictly speaking it should be related to the particular parameter θ entering into the differentiation.

Invariance. This term is mostly used to denote a property that does not get changed by a particular transformation. For example, an orthogonal transformation of a set of independent normal variates leaves the properties of independence and normality unaffected ; they have been invariant under the transformation.

Invariance Method. Refers to a principle of estimation or hypothesis testing which needs an estimator or hypothesis under test to remain invariant if the data of the problem undergo some transformation.

Invariance Principle. Refers to an alternative name for Donsker's theorem.

Inverse Correlation. See correlation coefficient.

Inverse Distribution. This expression has been used in several senses, for example :

(*i*) In a sense to be avoided, to denote the distribution of sample numbers where sampling gets continued until a predetermined number of successes is attained, *e.g.*, the 'inverse hypergeometric' has been the distribution of sample sizes n required to obtain k successes in sampling without replacement from a finite population.

(*ii*) an equally regrettable sense, to denote distributions (mainly discrete) where the frequencies have been reciprocal quantities, see for example, Waring distribution.

(*iii*) instead of expressing the distribution function $F(x)$ in terms of the variable x, to denote expansions of x in terms of $F(x)$.

It has been probably safer to avoid the expression altogether.

Inverse Moment. If the rth moment may be defined as $\xi(\chi_r)$, then the rth inverse moment may be defined as $\xi(\chi^{-}{}_r)$. A similar relationship is existing between the *r*th factorial moment and its inverse.

Inverse Multinomial Sampling. Refers to a form of sequential analysis in which sampling gets terminated when one or other of a specified subset of classes has occurred *k* times.

Inverse Probability. Refers to the probability approach which endeavours to reason from observed events to the probabilities of the hypotheses which could explain them, as distinct from direct probability, which reasons deductively from given probabilities to the probabilities of contingent events. A principal theorem in this connection has been Bayes' theorem. The word 'inverse' usually inverse in some logical relationship but is sometimes used in the sense of 'prier in time'. For example, in connection with the analysis of stochastic processes, it has been sometimes desirable to consider the past history of a system rather than its future development. The past probabilities of transition which have given rise to the present state have been thus sometimes called 'inverse'. This usage has been comparatively rate and not to be recommended.

Inverse Sampling. Refers to a method of sampling which needs that drawings at random shall be continued until certain specified conditions dependent on the results of those drawings are fulfilled.

Inverse Serial Correlation. A name sometimes, used for denoting the process of attempting to determine from a set of serial correlations the series which generated them. The problem has been in fact indeterminate as different series may give the same auto-correlations.

Inversion. If we define a standard order of elements (*e.g.*, the natural order of numbers 1, 2, 3, 4, 5,...), as inversion of two elements means that they are in the opposite order to that given in standard order. Thus, compared to 1, 2, 3, the order 1, 3, 2 contains one inversion (3, 2 instead of 2, 3). The concept has been used in the calculation of kendall's tau.

Investment Criteria. Refers to the several techniques of converting the yield of an asset to a capital sum, or the capital outlay to an annual cost, so that they can be added together or deducted one from the other ; also the several different ways of expressing the net yield, more precisely called the yield criteria.

Irreducible Markov Chain. A Markov chain is considered to be irreducible if all pairs of states of the chain communicate, so that the chain is having exactly one communicating class.

Irregular Variation. Refers to that variation within a time series which has been not accounted for by trend, seasonal components or cyclic components.

Ising-Stevens Distribution. If there have been n_1 and n_2 objects of two kinds C_1 and C_2 arranged at random in $n=n_1+n_2$ positions along a circle then the distribution (Stevens, 1939; Ising, 1925) of the number of runs of either object would be as follows :

$$\Pr(x)=\binom{n_1}{x}\binom{n_2-1}{x-1}\Big/\binom{n_1+n_2-1}{n_2}$$

$$=\binom{n_1-1}{x-1}\binom{n_2}{x}\Big/\binom{n_1+n_2-1}{n_1}$$

Isodynes. When contour maps of power have been plotted on a β_1, β_2 diagram, where β_1 and β_2 have been Pearson measures of skewness and kurtosis, lines of constant power are termed as isodynes.

Isometric Chart. Refers to a chart which tries to depict three-dimensional material on a plane. It has been a form of axonometric chart where the distances on the three axes have been measured on an equal scale. There have been various conventional combinations for the angles at which two out of three axes have been drawn *vis-a-vis* the horizontal.

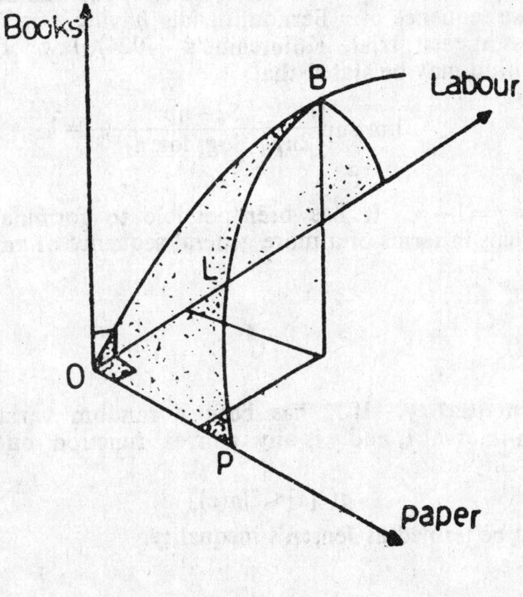

Example. A common use of isometric charts has been in the economics of production, where the two dimensions comprising the 'floor' of the box may be employed to illustrate two inputs, such as manhours and materials, and the vertical axis has been used to illustrate the volume of output of one product. The following isometric chart reveals levels of production of a printer, whose two major inputs have been labour and paper, and whose one major output is measured in quantities of books of a standard size. The shading illustrates the position where L units of labour and P units of papar have been used to manufacture B units of books.

Isomorphism. Refers to the logical equivalence of two theories, in the sense that one theory could be obtained from the other by a translation or reinterpretation of basic notions and symbols.

Isotropis Distribution. Refers to the distribution of a directional random variable that is having no preferred direction, *i,e.*, that is having a uniform distribution. By contrast, if the random variable does have a preferred direction, it is considered to have an anisotropic distribution.

Isotropy. A contingency table has been isotropic when the associations in all tetrads of any four frequencies for two rows and two columns have been of the same sign.

Iterated Logarithm, Law of. If S_n denotes the number of successes in a sequence of n Bernoulli trials having a probability p of success at each trial, Khintchine's (1924) law of the iterated logarithm may be stated that

$$\limsup_{n \to x} \frac{S_n - np}{2npq\,(\log, \log n)^{1/2}} = 1$$

where $q=1-p$. It has been possible to formulate stronger theorems in terms of a more general sequence of variates.

J

Jensen's Inequality. If X has been a random variable defined on an interval I, and g is any convex function on I, then the inequality

$$g(\xi[x] \leqslant \xi[g(x)]$$

would be termed as Jensen's inequality.

Jittered Sampling. This term is sometimes used in the analysis of time series (Parzen, 1967) for denoting the sampling of a continuous series where the intervals between points of observation have been the values of a random variable.

John's Cyclic Incomplete Block Designs. Refers to a form of incomplete block experiment which was proposed by John (1966) where the *t* treatments in blocks of *k* treatments per block could be divided into *k* cyclic sets of *t* blocks each.

Johnson's System. Refers to a system of frequency curves which is upon transformations of the variable. The original designation by Edgeworth (1898) had been the Method of Translation. The idea was extended by Johnson (1949) and developed in later paper.

Joint Cost. The cost of inputs that are inseparably attributable to two or more products ; the indivisible part of the cost of joint products, such as coal-gas, gas-coke and sulphate of ammonia ; beefsteak and sirloin ; petrol and diesel oil.

Joint Demand. Said of products and services that are in demand one with another. Petrol for motor vehicles, lubricating oils and tyres are said to be in joint demand ; so, too, are bread and butter. The joint demand for petrol, oil and tyres for cars is derived from the demand for travel, which is descrirbed as being in joint demand with petrol, etc. Petrol, etc. is in an indirect joint demand for travel.

Joint Distribution. Refers to the distribution of two or more variates. The term has been equivalent to multivariate distribution and has been especially used of two variates.

Joint Probability. The probability of two events occurring together or in succession.

Joint Regression. In the classical regression model it is assumed that the dependent variate has been a function, often linear, of a set of independent variables. If the linear model has been indequate, attempts have been sometimes made to allow for cross product terms in the independent variables, *e.g.*, if there have been two independent variables x_1 and x_2 the joint regression equation would be :

$$y=a_1x_1+a_2x_2+a_3(x_1x_2)$$

where the third term allows for any interaction between the original variables. Such regression has been sometimes said to be 'joint' but the usage has not seen to be very desirable.

Joint Sufficiency. Estimators $t_1, t_2, \ldots, t_k$ are considered to be jointly sufficient for parameters $\theta_1, \theta_2, \ldots, \theta_t$ if the likelihood function could be expressed as :

$$L(x_1 \ldots x_n ; \theta_1 \ldots \theta_t)$$
$$= L_1(t_1 \ldots, t_k ; \theta_1, \ldots \theta_t) \, L_2(x_1, \ldots, x_n)$$

where L_2 is not dependent on the $\theta_1 \ldots \theta_t$, although it may depend on other parameters of the system.

J-Shaped Distribution. Refers to an extremely asymmetrical frequency distribution having the maximum frequency at the initial (or final) frequency group and a declining or increasing frequency elsewhere. The shape of this distribution roughly resembles the letter J or its reverse. Among theoretical curves referred to as J-shaped have been the Pareto and certain of the Pearson system of frequency curves.

Judgment Sample. Refers to any sample which is not a probability sample. The dichotomy is not, perhaps, perfect ; for example, the visible stars of the sky are not a 'probability sample' of the matter in the universe, nor does it appear proper to describe them as a 'judgment sample'. It seems best to confine this term to the case where some element of human judgment is entering directly into the selection of the sample.

Judgment Sampling/Purposive Sampling. Any method of sampling where instead of using probability sampling the investigator uses his own 'expert' judgement to select elements which he considers to be typical or representative of the population. The problem with such methods is that there is no guarantee that the expert will not introduce serious biases into the results by basing the selection on faulty premises. In general, probability sampling is to be preferred.

Jump Matrix. For a Markov chain in continuous time, the matrix whose (i, j) element has been the conditional probability of a jump from state i to state j, given that a jump from state i occurs.

Jump Statistic. Refers to a statistic which is used to assess the reality of an apparent jump in the mean of a stationary time series.

Just Identified Model. A term which has been sometimes used to indicate a model which has been completely identifiable, but not overidentified.

K

Kagan-Linnik-Rao Theorem. This theorem (1965), which gets characterised the normal distribution by the property of constancy of regression of the sample mean on the deviations from this mean of the sample observations, states that if $X_1, X_2, \ldots X_n$, $n \geqslant 3$, have been independent and identically distributed random variables with finite expectation $\xi X_i = 0$, and

$$\xi(\bar{X} \mid X_1 - \bar{X}, \ldots, X_n - \bar{X}) = 0$$

where $\bar{X} = \frac{1}{n} \sum_{i=1}^{n} X_i$, then the X_i have been Normally distributed.

Kamat's Test. Refers to a nonparametric test for the equality of the scale parameters of two populations known to be having equal medians.

Kantorowitch's Theorem. According to this theorem, if X and Y be discrete random variables, X taking on N different values $x_1, x_2, \ldots x$N and Y taking on M different values $y_1, y_2, \ldots y$M, then if the relation

$$\xi(X^n Y^m) = \xi\,(X^n)\,\xi\,(Y^m), \quad 1 \leqslant n \leqslant N-1, \quad 1 \leqslant m \leqslant M-1$$

holds, then X and Y would be independent.

Kapteyn's Distribution. Refers to a generalisation of the Normal distribution using the method of Kapteyn's transformation of the form

$$d\phi_x = \frac{1}{\sigma\sqrt{(2\pi)}} \exp\left\{-\frac{1}{2\sigma^2}[G\,(x) - \mu]^2\right\} \left| \frac{dG(x)}{dx} \right| dx.$$

If G $(x) = x$ this gets reduced to the Normal distribution.

Karber's Method. Refers to a method which is used for estimating the median effective dose of a stimulus from data on quantal responses.

K-Class Estimator. In econometric regression analysis, this term refers to a general class of estimators which were proposed by Theil (1958) for the parameters in $By + \lambda z = u$ where y are the endogenous variables, z are exogenous and u are random errors. The method proceeds by least squares in stages.

Kendall's Coefficient of Concordance. Refers to a measure of the level of agreement (concordance) between several sets of matched results. Symbol : W. It takes values between 0 and 1. If there have been only two judges, a value of W near 0 implies disagreement between them ; if there have been many, a value near 0 means randomness in their judgments. A value near 1 reveals agreement between the judges.

Kendall's 'S' Score. In rank correlation analysis, Kendall (1955) defined a score, for a pair of rankings of n items, as $+1$ if any two have been ranked in the same order by the two rankings, -1 if in opposite order and zero if tied in either or both rankings. The total score S has been the algebraic sum of the $\frac{1}{2}n(n-1)$ contributions from pairs of items.

Kendall' Tau (τ). Refers to a coefficient of rank correlation which is based on the number of inversions in one ranking as compared with another.

Kendall's Terminology. Refers to a system of describing particular cases in queueing theory which was developed by D.G. Kendall (1953). It has been based upon initial letters of key words associated with the input, queue discipline and service mechanism of a congestion situation. For example, a random, Poissonian or Markov input may be represented by M and the number of service channels as 1 or $s(=1, 2,...)$.

Kesten's Process. Refers to a variation of the Robbins-Munro stochastic approximation process which is intended to reduce the bias related to the starting point. The step constant gets reduced only when there has been evidence that doses are stradling the median lethal dose.

Khintchine's Theorem. If $x_1, x_2,...$ denote a sequance of independent and identically distributed variates each with a finite mean

μ, then the variable $\bar{x} = \frac{1}{n} \sum_{i=1}^{n} x_i$ converges in probability to

μ as n tends to infinity. This theorem had been first proved rigorously by Khintchine in 1929.

Kollectiv. This word denotes a population of objects in which each unit bears one of a finite number of identifiable characteristics.

Kolmogorov Axioms. Refers to a set of axioms which were given by Kolmogorov (1933) for the foundation of probability theory in terms of set-and measure theory. They have been forming the starting point of most modern expositions of a mathematical theory of probability.

Kolmogorov Inequality. Refers to a generalisation of the Bienayme-Tchebychev inequality. If $x_i (i=1, 2, ..., n)$ be n mutually independent variates with means a_i and variances v_i and if $A_k=\sum_{i=1}^{k} a_i$ and $V_k=\sum_{i=1}^{k} v_i$, and let $S_k=\sum_{i=1}^{k} x_i$, then for $t>0$ the probability that the inequalities

$$|S_k-A_k| < tV_k$$

have been simultaneously realised has been at least equal to $1-1/t^2$.

Kolmogorov Representation. Refers to a representation of the logarithm of the characteristic function $\theta(t)$ of infinitely divisible distributions having finite second moment. The representation is having the form

$$\log \theta(t) = i\beta t + \left[\frac{e^{itu}-1-itu}{u^2}\right] dK(u)$$

where β is a real constant, and the function K coincides, to within a non-negative multiplicative constant, with a distribution function.

Kolmogorov-Smirnov Distributions. Refer to a name assigned to a variety of distributions which have been exact or approximate distributions of functions of the sample cumulative distribution function.

Kolmogorov-Smirnov Test. Is a non-parametric test, which does not require that data be grouped in any way, for determining whether there is a significant difference between an observed frequency distribution and a theoretical frequency distribution.

Kolmogorov Theorem. Refer to two theorems which were stated by Kolmogorov (1928-30) and give the conditions under which the strong law of large numbers holds :

(*i*) if the random variables are independent ;

(*ii*) if the random variables are independent and identically distributed.

Konyus Index Number. Refers to the name which is assigned to a class of index number rather than to one specific formula. It is an index of priees based on quantities, the 'budget' of the consumer, which are optimal in some field of consumer preference. If the indifference level for consumers has been specified in terms of an optimal budget at the base prices, the index is termed as Laspeyres-Konyus ; if by an optimal budget at prices in the period under comparison with the base period, a Paasche-Konyus index.

Koroljuk's Theorem. A theorem given by Koroljuk, which states that for a regular stationary point process $\{N(t), t \geqq 0\}$, the mean number of events per unit time, μ, and the intensity λ defined by

$$\lambda = \lim_{t \to 0} \frac{1 - \Pr(N(t)=0)}{t}$$

have been identical, including the case where both λ and μ have been infinite.

Kronecker Product of Design. It was concluded that an experiment design D uniquely as certains its incidence matrix and vice versa. Thus the Kronecker product (in the sense of matrix theory) of two incidence matrices produces a new design matrix and can be used to as certain the existence of certain classes of design.

Kronecker Product of Matrices. The product $A \times B$ of an $m \times m$ matrix A and an $n \times n$ matrix B has been the $mn \times mn$ matrix whose elements have been the products of terms, one from A and one from B.

Kruskal Statistic. It is a rank order statistic for the k-sample problem.

$$H = \{12/N(N+1)\} \Sigma n_i \{\bar{R}_i - (N+1)/2\}^2$$

where $i = 1, 2, \ldots, k$, n_i denote the number in the ith sample, N the total Σn_i and R_i the average rank sum in the ith ranking.

Kruskal-Wallis One-Way Analysis of Variance. A non-parametric test of the null hypothesis that three or more samples are drawn from the same parent population. Symbol : H. The test has been performed on the overall ranks of the items sampled, and so has been a test for differences in location of the particular samples.

Kruskal-Wallis Test. Refers to a multi-sample rank-randomisation test for identical populations, sensitive to unequal locations. It has been a direct generalisation of the two-sample Wilcoxon rank sum test to the multi-sample case.

K-Samples Problem. Refers to the problem of determining whether, given k samples, one from each of k populations, the parent populations have been different. The usual tests developed in this connection have been homogeneity tests for means or variances, but an infinite number of tests have been possible.

K-Statistics. Refers to a set of symmetric functions of sample values which were proposed by R.A. Fisher (1928). The univariate k-statistic of order r may be defined as the statistic whose mean value has been the rth cumulant, k_r, of the parent population. The statistics have semi-invariant properties and their sampling cumulants can be got directly by combinatorial methods.

Similarly the multivariate k-statistic, say $k(r, s, \ldots, v)$ may be defined as the symmetric function of observations whose mean value is the corresponding cumulant $k_{rs \ldots v}$. It is more usually written $K_{rs \ldots v}$.

Another generalisation has been due to Tukey, who defines a statistic, say $k\{r, s, \ldots, v\}$ as the symmetric function whose mean value is $k_r k_s \ldots k_v$. This statistic has been also often written $k_{rs \ldots}$, and should not get confused with the multivariate k-statistic.

KS Statistics. Said to be short for *keep-in-step statistics*, introduced during World War II in UK for economic control purposes, but found to be so useful to industry and the Government that their collection and collation have been continued. They consist, for the most part, of sales and delivery figures for a wide variety of products, on a monthly or quarterly basis.

K-Test. Refers to a distribution-free test for a trend in a series which was proposed by Mann (1945). If the equally spaced series has been $x_1, \ldots, x_n$ and a decreasing trend exists each term will tend to be greater than succeeding terms. The smallest interval K for which $x_i < x_{i+k}$, $i = 1, 2, \ldots, n-K$ has been taken as the test statistic. the null hypothesis that no trend exists being rejected if K has been small. A similar test, of course, exists for increasing trend.

Kuder Richardson Formula. A formula which is used for estimating the reliability coefficient of a test which has been endeavoured to overcome the disadvantages of formulae associated with split half methods. There have been numerous versions of the formula, which may be written

$$r_{tt} = \left(\frac{n}{n-1}\right)\left(\frac{\sigma_t^2 - \Sigma pq}{\sigma_t^2}\right)$$

where n denotes the number of items in the test, p the proportion passing or satisfactorily responding to an item, $q = 1 - p$ and σ_t^2 the overall variance of scores.

Kuiper Statistic. It is a goodness-of-fit test statistic which is similar to the Kolmogorov-Smirnov statistic. For the single-sample problem it is

$$V_n = n^{1/2}[\text{Sup}_x\{F_n(x) - F(x)\} - \text{Inf}_x\{F_n(x) - F(x)\}]$$

and for the two-sample problem

$$V_{m,n} = [mn/(m+n)]^{1/2}[\text{Sup}_x\{F_n(x) - G_m(x)\} - \text{Inf}_x\{F_n(x) - G_m(x)\}].$$

Kurtosis. The sharpness of a peak on a curve of a probability density function (see Fig.). A distribution with high kurtosis is termed as *leptokurtic* ; one with medium kurtosis, *mesokurtic*, and one with low kurtosis, *platykurtic*.

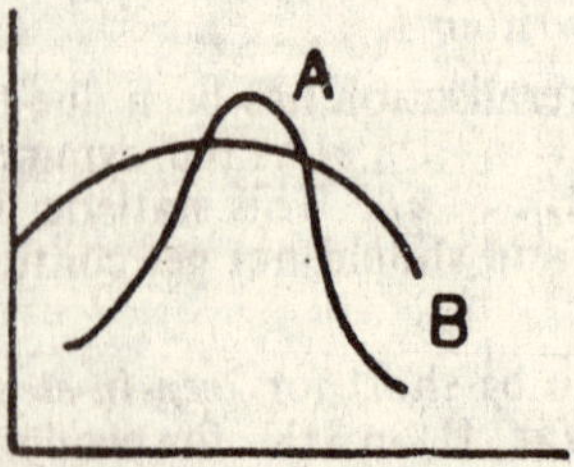

Fig. Kurtosis. Distribution A has larger kurtosis than B.

A measure of kurtosis has been the *coefficient of kurtosis*, which has been defined as :

$$\frac{E[(X-\mu)^4]}{\sigma^4} - 3$$

If a curve is sharply peaked at the centre, the curve is leptokurtic. If the central section of the curve is flattened or rounded it is platykurtic. An intermediate degree of peakedness such as that of the normal distribution is called mesokurtic. The accompanying diagram illustrates the three curves.

A curve showing the variation in the kurtosis of one variate against values of the other in bivariate variation is called a kurtic curve.

L

Labour-Time Function. Man hours or man days expressed as a function of work done ; usually of linear form, and not to be confused with the employment function, which is of logarithmic form.

Labour Turnover. The ratio of the number of personnel leaving the firm, in unit time, to the total number employed. Unit time is usually a calendar month, and the number employed is the average during the month, the average being, say, that of the number employed on the first and last working days and the working day nearest the middle of the month.

Ladder Indices. A ladder index may be defined as the time point (epoch) in a random walk at which a ladder variable begins to take place.

Ladder Variable. If, in a random walk, the sum of the mutually independent variables at time point (epoch) n is S_n then, if $S_n > S_j$ ($j=0, 1, 2, \ldots, n-1$), the variable $\bar{X}_n$ has been an ascending ladder variable. Descending ladder variables are similarly defined where $S_n < S_j$.

Lag. An event occurring at time $t+k (k>0)$ is considered to lag behind event occurring at time t, the extent of the lag being k. An event occurring k time units before another may be considered as having a negative lag.

Lag Correlation. Refers to the correlation between two series where one of the series is having a lag with reference to the other.

Lag Covariance. Refers to the first product moment between two series, one of which gets lagged in relation to the other. For example, if u_t and v_t have been two series defined at $t=1, 2, \ldots, n$, the lag covariance of u_t and v_t of order $k>0$ would be as follows :

$$\frac{1}{n-k}\sum_{1}^{n-k} (u_t - \bar{u}_1)(v_{t+k} - \bar{v}_2)$$

where

$$\bar{u}_1 = \frac{1}{n-k}\sum_{1}^{n-k} u_t \quad \text{and} \quad \bar{v}_2 = \frac{1}{n-k}\sum_{1}^{n-k} v_{t+k}.$$

Lag Hysteresis. Lag hysteresis in econometrics has been referred to cases of sinusoidal variation in the two variables : skew hysteresis is referred to the more realistic case in which the oscillatory movements are asymmetrical. The value of the concept where oscillatary movements have been not cyclical has been unknown.

Lag Regression. Refers to a regression in which the values of the dependent variate and one at least of the independent variables get lagged in relation to each other.

Laguerre Polynomials. These are the polynomials which are due to Laguerre and defined by

$$L_n(x)=e^x D^n(x^n e^{-n})/n!$$

where $D=d/dx$. They are having important orthogonal properties. The polynomials up to the third order are : $L_0(x)=1$; $L_1(x)=1-x$; $L_2(x)=(x^2=4x+2)/2!$; $L_3(x)=(-x^3+9x^2-18x+6)/3!$

Lancaster's Partition of Chi-Squares. A total chi-square value has been computed to provide a gross measure of the extent to which the cell frequencies get departed from expectation. This value has been partitioned into additive components to show how much has been attributable to individual classifications and how much is due to both first order and second order interaction. This procedure can be extended to cover contingency tables involving more than three ways of classification where higher order interactions occur. A special merit of this procedure has been that it can be used with theoretical parameters or with parameters estimated from the data.

Laplace Distribution. Refers to a frequency distribution of the double exponential type which is expressible in the form

$$dF=\frac{1}{2\sigma}\exp\left\{-\frac{|x-m|}{\sigma}\right\}dx,\ -\infty\leqslant x\leqslant\infty\ ;\ \sigma>0.$$

It is sometimes termed as the first law of Laplace, in contradistinction to the second law which is the same as the Normal distribution.

Laplace's Theorem. This limit theorem, of which the Bernoulli theorem has been a corollary, may be stated that if there are n independent trials, in each of which the probability of an event is p, and if this event occurs k times, then

$$\Pr\left\{z_1\leqslant\frac{k-np}{\sqrt{(npq)}}<z_2\right\}\to\frac{1}{\sqrt{(2\pi)}}\Big]_{z_1}^{z_2}e-\tfrac{1}{2}z^2\,dz$$

as $n\to\infty$ whatever the numbers z_1 and z_2.

Laplace Transform. If a function $g(t)$ has been related to a second function $f(x)$ by the equation

$$g(t)=\int_0^\infty e^{-tx} f(x)\, dx$$

then $g(t)$ has been the Laplace transform of $f(x)$.

In statistical theory it has been more customary to use the Fourier transform, which has certain advantages over the Laplace form ; *e.g.*, if $f(x)$ has been a frequency function the Fourier transform always exists whereas the Laplace transform may not do so for real t.

Large Numbers, Law of. A general form of this fundamental law relating to random variables could be stated as follows :

If x_k denotes a sequence of mutually independent variates with a common distribution and if the expectation $\mu=\xi(x_k)$ exists, then for every $\varepsilon>0$ as $n\to\infty$ the probability

$$\Pr\left\{\left|\frac{X_1+\dots+X_n}{n}-\mu\right|>\epsilon\right\}\to 0.$$

Laspeyres' Index. A form of index number due to Laspeyres (1871). If the prices of a set of commodities in a base period are p_o, p_o', p_o'',...and those in a given period p_n, p_n', p_n'',...; and if q_o, q_o', q_o'' are the quantities sold in the base period, the Laspeyres' price index number is written

$$I_{on}=\frac{\Sigma\,(p_n\, q_o)}{\Sigma\,(p_o\, q_o)}$$

where the summation takes place over commodities. In short, the prices are weighted by quantities in the base period.

Generally, an index number of the above form is said to be of the Laspeyres' type even when p and q do not relate to prices or quantities ; the characteristic feature being that the weights relate to the base period, as contrasted with the Paasche index in which they relate to the given period.

Laspeyres Method. To weight an aggregates index, this method uses as weights the quantities consumed during the base period.

Latent Structure. In general this term refers to a structure expressed in terms of variates or variables which have been 'latent' in the sense of not being directly observable. Certain econometric relations *e.g.*, in terms of 'utility' have been of this type, and the models used in factor analysis as used in psychology also may be considered as a kind of latent structure.

Latent Variable. A variable which has been unobservable but has been supposed to enter into the structure of a system under study, such as demand in economics or the 'general' factor in psychology, Unobservable quantities such as errors have been not usually described as latent.

Latin Cube. Refers to an extension of the principle of the latin square so that the layers, and sections in two directions, have been effectively latin squares.

Latin Rectangle. An experimental design which has been derived from the latin square. It consists of a latin square with one or more adjacent rows or columns added or omitted. This particular design has been one form of the incomplete latin square or Youden square.

Latin Square, Latin Square Design. A term used in the design of experiments for a design which attempts to remove the experimental error resulting from the variation from two sources which can be identified with the rows and columns of the square by permutation of the treatments, each of which is identified with a Latin letter.

Lattice Diagram. In quality control, it is a diagram on which the results of attribute tests on samples are illustrated.

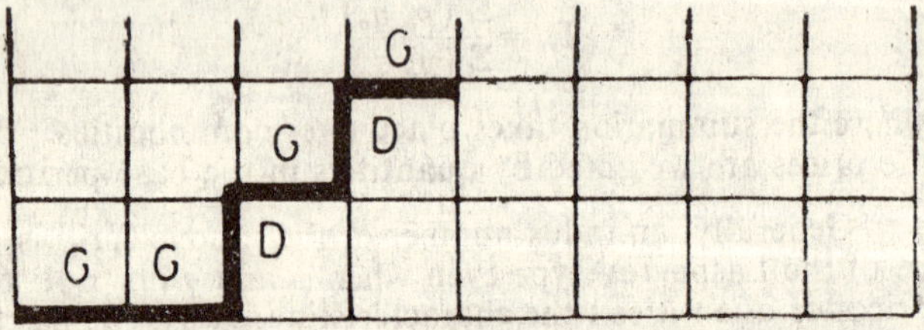

Fig. 1. Lattice diagram. This sampling line illustrates whether the first six items to be tested were good G or defective D in a quality-control test.

Lattice Distributions. Refer to a class of distribution, which includes most of the discrete distributions used in statistics, in which the intervals between any one random variable, for which there have been non-zero probabilities, have been integral multiples of one quantity (which depends on the random variable) and thus form a lattice. By suitable transformations all variables could be made to take values which are integers.

Lattice Sampling. Refers to a method of sampling in which substrata have been selected, for the sampling of individuals, according to some pattern analogous to the allocation of treatments on a lattice experimental design.

Laurent Process. A stochastic process for which the covariance generating function has been a Laurent series ; the important case has been that in which the process has been generated by a moving average of a random process.

Law of Error. States that the frequencies with which errors of measurement and differences between actual figures and estimates occur tend to form a symmetrical distribution curve which, in terms of standard deviations, conforms to the normal curve of error ; sometimes called the *normal probability curve* or the *normal-frequency-distribution curve.*

Law of Large Numbers. The law is a popular interpretation of the principle of statistical regularity, and states that large groups or aggregates of data show a greater degree of stability than small ones. Thus if a reasonably large sample of items is drawn at random from a parent group (or population) it will probably show the characteristics of the parent group more representatively than a small sample would do. In fact, the sampling error is inversely proportionate to the square root of the number of items in the sample.

Learning Curve. A curve which describes the upward trend in productivity, or the downward trend in labour requirements, per unit of output as the employees gain experience in carrying out particular processes or in producing certain goods.

Least Favourable Distribution. Given the composite hypothesis $H_0 : f\theta,\ \theta\epsilon\omega$, to be tested against the simple alternative H, the general procedure has been to reduce the composite hypothesis H_0 to a simple one $H_0\lambda$, where λ is a probability distribution function over ω. The maximum power β_2 that can be obtained against H_1 has been that of the most powerful test of $H_0\lambda$ against H_1. The probability distribution λ has been said to be least favourable (for a given level α) if, for all λ' the inequality $\beta\lambda \leqslant \beta\lambda'$ holds.

Least Significant Difference Test. Refers to a test for comparing mean values arising in analysis of variance. It has been an extension of the standard *t*-test for the difference between two specified mean values. As the tests between pairs are not independent, the error rate is difficult to assess exactly.

Least Squares Estimate. An estimate obtained by a least squares process or method. For example, in least squares regression analysis a line of best fit is obtained and the relationship between observations of a dependent variable and the regressors is expressed as an equation. The exact points on the line corresponding to the observations ate estimates, not in the usual statistical sense (*e.g.*, an estimate of a parameter from a statistic) but in the more general sense, that they are values produced by a rule of estimation (in this case the least squares regression model or equation).

Least Squares Estimator. In statistical analysis an estimator is a rule by means of which the parameters of a parent group may be estimated, the values produced being termed estimates. In any technique using least squares method (as, for example, least squares regression analysis), the estimator is the rule or model obtained by least squares (for example, that y should be estimated as $a+bx$) ; and the estimate is the value produced by the rule.

Least Squares Method. A method of estimating a function or line of best fit by employing the principle that the optimal function is test which minimizes the squares of differences between the estimates which it produces and the observations on which it is based.

Least Variance Difference Method. It is a method of gereralised linear estimation of structural parameters.

Legendre Polynomials. Refers to a set of polynomials which are due to Legendre (1785). They have been the coefficients of $P_n(x)$ in the expansion

$$(1-2xh-h^2)^{-(1/2)}=\sum_{n=0}^{\infty} P_n(x)h^n.$$

In particular

$$P_0(x)=1,\ P_1(x)=x,$$

$$P_2(x)=\tfrac{1}{2}(3x^2-1),\ P_3(x)=\tfrac{1}{2}(5x^3-3x).$$

They are having important orthogonal properties and can get transformed linearly so that they have been orthonormal on the interval (0, 1).

Legit. Refers to a transform of quantal data which are used in genetics. A proportion p, regarded as a gene ratio, in certain circumstances gets connected with a variate x by the differential equation.

$$\frac{\partial^2 p}{\partial x^2}=4pqx$$

where $q=1-p$, and with the boundary conditions that $p=\tfrac{1}{2}$, $x=0$ and $p\to 0$, $x\to\infty$. x as a function of p is called a legit.

Leptokurtic. A strongly peaked distribution.

Leptokurtosis. See kurtosis.

Leslie's Test. In capture-recapture experiments, the probabilities of capturing each living member of the population must be the same. In order to test this assumption, Leslie (1958) proposed the statistic

$$L = \sum_{i=1}^{N} (R_i - \bar{R})^2 \Big/ \sum_{g=f}^{g} p_j q$$

where R_i = the number of recaptures of individual i between the fth and gth sample inclusive ; $\bar{R} = \sum_{i=1}^{n} R_i/N$ and P_j is the proportion of recaptures at the jth sample. The statistic has been distributed as χ^2 with N−1 *d.f.*, although it was later shown by Carothers (1971) that $(1-1/N)L$ has been identical to Cochran's Q-test.

Level Map. A graph which shows curves in an (x, y) plane corresponding to various values of the constant k in a defining equation $f(x, y) = k$. It has been similar to the contours of equal height on a geographical map.

Level of Measurement. A term used to denote the strength of the scale of measurement on which the data are measured. The lowest level of measurement is the nominal scale for which the only applicable statistical methods are those based on frequencies or counting the mode, contingency tables, etc.

Level of Significance. There are many statistical tests of hypotheses which depend on the use of the probability distributions of a statistic t chosen for the purpose of the particular test. When the hypothesis has been true this distribution has a known form, at least approximately, and the probability $\Pr(t \geqslant t_1)$ or $\Pr(t \leqslant t_0)$ can be determined for assigned t_0 or t_1. The acceptability of the hypothesis has been usually discussed, *inter alia*, in terms of the values of t observed ; if they have a small probability, in the sense of falling outside the range t_0 to t_1, $\Pr(t \geqslant t_1)$ and $\Pr(t \leqslant t_0)$ small, the hypothesis is rejected. The probabilities $\Pr(t \geqslant t_1)$ and $\Pr(t \leqslant t_0)$ are called levels of significance and are usually expressed as percentages, *e.g.*, 5 per cent. The actual values are, of course, arbitrary, but popular values are 5, 1 and 0.1 per cent.

Levy-Khinchine Representation. Refers to a representation of the logarithm of the characteristic function $\phi(t)$ of an arbitrary infinitely divisible distribution. The representation is having the form

$$\log \theta(t) = i\beta t + \int \left(e^{itu} - 1 - \frac{itu}{+u^2} \right) \times \frac{1+\mu^2}{u^2} \, dG(u)$$

where β is real and G coincides with a distribution function to within a non-negative multiplicative constant.

Levy-Pareto Distribution. The work of Levy in connection with stable distributions proved that the Pareto distribution ,could be derived from a version of the central limit theorem in which the individual random variables do not have finitc variance.

Levy Representation. Refers to a representation of the logarithm of the characteristic function $\theta(t)$ of an arbitrary infinitely divisible distribution. The representation is having the form

$$\log \theta(t) = i\beta t - \sigma^2 t^2/2 + \int_0^{\infty} h(t, u) dN(u) + \int_{-\infty}^{0} h(t, u) dM(u)$$

where β is real, $\sigma > 0$, $h(t, u) = e^{itu} - 1 - itu/(1+u^2)$ and the functions M and N are able to satisfy a number of conditions.

Lexis Ratio. This ratio is regarded as a measure of distinguishing the three kinds of variation in sampling for attributes : Bernoullian, Lexian and Poissonian.

If k samples of $n_1, n_2, \ldots, n_k$ members bear observed proportions of the attribute $p_1, p_2, \ldots, p_k$ the Lexis Ratio Q may by defined by

$$Q^2 = \frac{\sum_{j=1}^{k} n_j (p_j - p)^2}{(k-1)pq}$$

where p denotes the proportion of the attribute in all samples together and $q = 1 - p$. If the Lexis ratio has been equal to unity within sampling limits the sampling has been regarded as Bernoullian : if greater than unity, as of Lexian type ; if less than unity, as Poissonian. They are also said to have normal, hypernormal (or supernormal) and subnormal dispersion. In Italian usage the terms hyperbinomial, binomial and hypobinomial have been sometimes employed.

Liapounov's Inequality. An inequality which had been due to Liapounov (1901). It is concerning the relations between the absolute moments of a frequency distribution. If $a \geqslant b \geqslant c \geqslant 0$ have been three real numbers and v_a, v_b, v_c the absolute moments of orders a, b and c for some arbitrary distribution, then :

$$v_b^{a-c} \leqslant v_c^{a-b} \, v_a^{b-c}$$

Liapounov's Theorem. Refers to a form of the central limit theorem which assumes the existence of absolute third moments. If x_j ($j=1, 2, \ldots, n$) has been a sequence of independent variates with means m_i, variances σ^2 and absolute third mean-moments ρ_j^3 the sum $\sum_{j=1}^{n} x_j$ has been asymptotically Normal provided that $\lim_{n \to \infty} \rho/\sigma = 0$

where

$$\rho^3 = \sum_{j=1}^{n} \rho_j^3 \text{ and } \sigma^2 = \sum_{j=1}^{n} \sigma_j^2.$$

Life Table. A table which shows the number of persons who, of a given number born or living at a specified age, live to attain successive higher ages, together with the numbers who die in the intervals.

Likelihood. A measure of the plausibility of a postulated population structure given values of observations on a random sample from that population. The postulated population structure is usually in terms of a model for the distribution function of a random variable defined on that population together with values for parameters of that function.

Likelihood Function. This is likelihood which is expressed as an algebraic function of possible values of parameters for a particular population distribution and the observed set of sample data.

Likelihood Ratio. Refers to the ratio of the likelihood of one postulated population structure to another for a given set of sample data. Likelihood ratios form the basis for likelihood ratio tests.

If $x_1, x_2, \ldots, x_n$ are a random sample from a population $f(x\,;\,\theta_1, \theta_2 \ldots \theta_k)$ the likelihood of this particular sample would be

$$L = \prod_{i=1}^{n} f(x_i, \theta_1, \theta_2, \ldots, \theta_k).$$

Likelihood Ratio Tests. These have been tests of null hypotheses of a population structure against specified alternative hypotheses which are based on the likelihood ratio. Usually the null hypothesis is simple in that it specifies the population distribution and values of parameters. The likelihood of the null hypothesis given the sample data can be known by substituting these values in the likelihood function. The alternative hypothesis may sometimes be simple in that it specifies particular values for the parameters.

Limited Information Methods. In econometrics, these refer to methods of deriving estimates of parameters in a stochastic system which are not using all the information available. The term is generally confined to those methods which give consistent estimates, *i.e.*, are unbiassed for large samples.

Lincoln Index. It is a method of estimating the size of populations of mobile units, *e.g.*, animal populations which relies on the capture, marking, release and recapture of samples from the population under investigation.

Lindeberg-Feller Theorem. Refers to a form of the central limit theorem which provides a necessary and sufficient condition for the distribution of a sum of independent random variables to be asymptotically normal.

Lindeberg-Levy Theorem. Refers to a particular case of the central limit theorem when all the variates concerned are having the same distribution.

Lindley's Theorem. Refers to a limit theorem in queueing theory which concerns the distribution function of the waiting time of the nth arriving customer.

Linear Constraint. Refers to a condition imposed on certain variate values or frequencies which has been linear in form. For example, if samples of variates $x_1, x_2, \ldots, x_n$ have been drawn, their mean, in general, will vary ; but if only those samples have been considered which have a mean equal to zero, all other samples being ignored, the variates have been subject to the linear constraint $\sum_{i=1}^{n} x_i = 0$. The distribution under constraint may be considered as conditional.

Linear Correlation. An absolete expression. It was used to denote either (*a*) the product moment correlation in cases where the corresponding regressions were linear or (*b*) a coefficient of correlation constructed from linear functions of the observations. The expression has been best avoided altogether.

Linear Discriminant Function. Refers to a discriminant function which is a linear function of observed variate.

Linear Estimator. An estimator which has been a linear function of the sample values. A linear estimator for the parent mean has been :

$$\frac{a_1x_1+a_2x_2+\ldots+a_nx_n}{a_1+a_2+\ldots+a_n}$$

where $x_1, x_2, \ldots, x_n$ have been the sample values and $a_1, a_2, \ldots, a_n$ all constants.

In the case when $a_1=a_2=\ldots a_n=1$, this becomes

$$\frac{x_1+x_2+\ldots+x_n}{n}$$

where is the sample mean, x.

Linear Hypothesis. Logically, this expression is used to relate to any statistical hypothesis concerning the parameters of a distribution which could be expressed linearly in terms of them. For example, with the distribution.

$$dF=e^{-(x-m)/\sigma}dx/\sigma,\ m\leqslant x\leqslant\infty$$

the hypotheses $m=m_0$ or $m-\sigma=0$ have been linear, whereas $m^2-\sigma=0$ is not liner, although it could be kept into a linear form by transforming the parameter m to m^2.

Linear Maximum Likelihood Method. Refers to a method of parameter estimation in multi-equation models in which the Taylor series linear approximations of the partial derivatives of the likelihood function get equated to zero rather than the actual partial derivatives.

Linear Model. Refers to a model in which the equations connecting the variates or variables have been in a linear from.

Linear Process. It is a stochastic process which is defined by the formal expression

$$x_t=\sum_{u=-\infty}^{t} g_{t-u}w_u$$

for the discontinuous case or the analogous integral

$$x(t)=\int_{-\infty}^{t} g(t-u)\ dw(u)$$

for the continuous case, where w_u or $dw(u)$ represent independent and stationary disturbances.

Linear Program (Linear Programming). It is a branch of mathematical programming which deals with making quantitative decisions when several sets of differing resources and alternative courses of action are involved in the decision process. In the case of linear programming all the functions defining objectives (ends) and constraints (restrictions or limitations of means) are linear. Linear programs employ iterative, stepwise procedures for determining the optimum solution in any particular case.

Linear Programming. A series of methods originating in operational research based on the assumption that relationships between factors are of a linear (straight-line) type. The methods include break-even analysis and optimizing product mix.

Linear Regression. Refers to a regression relationship between one or more independent variates (X_j) and a dependent variable (Y) where there has been an expectation that the dependent variable has been a first-degree function of the independent variate (s) (or regressor (s)).

Linear-Regression Analysis. Regression analysis, both simple and multiple, based on a linear model.

Linear Structural Relation. Refers to a relationship in linear form among the observed endogenous variables.

Linear Systematic Statistic. Refers to a systematic statistic which has been linear function of the observations. Most 'systematic' statistics in current use have been in fact, linear and the linearity has been sometimes taken as understood in describing the linear statistic simply as 'systematic'.

Linear Trend. Refers to a trend for which the value has been a linear function of the time variable, *e.g.*, $u(t)=a+bt$ where a and b have been constants.

Line of Best Fit. It is the line which fits a set of data most accurately when plotted on a graph. It depends on how greatest accuracy is defined in any example.

Line of Equal Distribution. It is the straight line on the graph of a Lorenz curve which is possing through the origin and the upper extreme of the curve ; when, as is customary, the two variables vary from 0 to 1, this has been the line making an angle of 45° with the coordinate axes. It provides a reference line of equal distribution, production, concentration, etc.

Link Relative. In index number theory, the value of a magnitude in a given period divided by the value in the previous period. It is trend ratio.

Line Sampling. Refers to a method of sampling in a geographical area. Lines have been drawn across the area and all members of the population falling on the line, or intersected by it, have been included in the sample. If the lines have been straight parallels equally spaced across the area concerned, then the sampling becomes one form of systematic sampling. If, instead of all intercepts on the lines, a series of evenly spaced points have been chosen on each line, the sampling has been equivalent to choosing the points on a lattice and may also be considered as two stage line sampling.

Linked Samples. Two samples of the same size in which there is a one-to-one correspondence between their respective sample units. The linking method may provide either an absolute or partial constraint on the selection insofar as the selection of a unit in one sample may determine uniquely the element to be selected in the other, or may simply restrict the choice of the corresponding element.

List Sample. Refers to a sample which is selected by taking entries from a list of the items constituting the population under review. The usual method of selecting entries has been to take them at equal intervals, the starting point being selected at random.

Local Statistic. Refers to a statistic, estimator or test statistic, which gets derived from short term comparisons within a time series.

Location Parameter. Refers to a parameter which 'locates' a frequency distribution in the sense of defining a central or typical value like a mean or mode.

Location Shift Alternative Hypothesis. If there is a random sample of observations from $K(\geq 2)$ populations, in testing the hypothesis that these populations are identical, the alternative hypothesis could be that the only difference between the parent distribution is one of location. This alternative hypothesis is termed as the location shift alternative hypothesis.

Log-Chi Squared Distribution. A transformed distribution which is used in the analyses of heterogeneous variances and the analyses of Poisson processes.

Logarithmic Chart. A graph whereon one or both axes have been scaled in terms of logarithms of the variables. The chart may be termed as a semi- or double logarithmic chart according to whether only the ordinate or both the ordinate and abscissa are on a logarithmic scale.

Logarithmic Normal (Lognormal) Distribution. If the logarithms of a set of variate values have been distributed according to the normal distribution the variate has been said to have a logarithmic normal distribution, or be distributed 'lognormally'.

Logarithmic Scale. A chart which is drawn to logarithmic scale in one or both of its axes uses the intervals and gradations of equivalent logarithmic distances instead of normal ordinate and co-ordinate measurements.

Logistic Distribution. Refers to a frequency distribution of the form $\beta\{e^t/(1+e^t)^2\}$ with $\beta>0$. It closely approximates the normal distribution and could also be shown to be the asymptotic-distribution of the mid-range of an exponential-type parent distribution. In its cumulative form it is also used as a growth curve.

Logistic Process. Refers to a stochastic process which is associated with the logistic law of growth. It has been a particular case of a birth and death process in which the rates for these two phenomena have been linearly dependent upon population size.

Logit. In some problems relating to the proportion of subjects responding to different doses of a stimulus the model based on a tolerance distribution has been not appropriate. A logistic relation

$$P=\{1+e^{-(\alpha+\beta x)}\}^{-1}$$

may approximately denote the dependence of probability of response P on dose x.

Loglog Transformation. Refers to the transformation of a probability P to a response metameter Y according to the following formula

$$Y=\log_e(-\log_e P).$$

Log Paper. There are two kinds of log paper : *log-log paper*, in which both axes have ratio scales, and *semi-log paper*, which has a vertical axis with a ratio scale, and a horizontal axis with a normal scale. A ratio scale is one on which any ratio is given equal distance on all sections of the scale

Plotted on semi-log paper, a geometric progression would describe a straight line, whereas on rectilinear graph paper, it would describe a curve.

Log-log paper is used by statistical analysts for plotting the statistical series of two factors, one against the other, where it is thought that there is a logarithmic relationship between the two. If the scatter of dots lies about a straight line, the hypothesis is probably true ; if it lies about a curve, or is too wide to suggest a graph of any shape, the hypothesis is probably false.

Lomax Distribution. A name which is sometimes given to an alternative form of the Pareto distribution. This term is also used to refer to as the Pareto distribution of the second kind.

Longitudinal Surveys, Longitudinal Studies, Panel Surveys. A longitudinal survey or panel survey is a survey in which information is sought from the same samples of respondents on more than one occasion.

Lorenz Curve. A graph which has been designed to show the extent of inequality in the distribution of a variable throughout a population. Percentiles of the population have been plotted against percentiles of the variable concerned.

Loss Function. In the making of decision on the basis of observations on a variate x, disadvantage could be suffered through ignorance of the true distribution of x. The extent of the disadvantage has been often a function of the true distribution and of the decision which is actually made. This function has been called the loss function.

Lot. A lot has been a group of units of a product produced under similar conditions and therefore, in a sense, of homogeneous origin ; *e.g.*, a set of screws produced by a lathe or a set of electric light bulbs produced by a number of similar machines.

Lottery Sampling. Refers to a method of drawing random samples from a population by constructing a miniature of the population, *e.g.*, by inscribing the particulars of each member on to a card and drawing members at random from it, *e.g.*, by shuffling the cards and dealing a set haphazardly.

Lower-Tailed Test. A one-tailed hypothesis test in which a sample value significantly below the hypothesized population value will lead us to reject the null hypothesis.

L-Statistics. Refers to a class of rank order statistics which was proposed by Dwass (1956) in connection with rank order tests in the two sample problem ; this must not get confused with the Neyman & Pearson L-tests for homogeneity of sample variances. The L-statistic may be defined as $\Sigma aN_i bNR_i$ where the aN_i have been $(n/mN)^{1/2} - i = 1,\ldots,m$, and $i = m+1,\ldots,N$ being the two samples under test—the bN another set of constants and $R_1 \ldots Rn$ the ranks of the N random variables.

L-Tests. Refer to the tests which are proposed by Neyman & E.S. Pearson (1933) for testing the homogeneity of a sat of sample variances. The tests, which have been based on likelihood ratios in normal variation, vary according to the precise type of hypothesis under test ; for example, if the hypothesis has been that the parent variances of k sample have been all equal but means may differ the test statistic is

$$L_1 = \prod_{i=1}^{k} (s_i^2)^{n_i/N} \Big/ \overline{N} \sum_{i=1}^{k} (n_i S_i^2)$$

where N represents the total number of observations, n_i the number of observations of the ith sample and s_i^2 the variance of the ith sample. Small values of the test function give rise to a rejection of the hypothesis.

M

Magic Square Design. Refers to a square array $(n \times n)$ with n^2 integers kept in the cells in such a way that the row column and two principal diagonal sums have been the same ; namely $\frac{1}{4}n\ (n^2+1)$. These squares may be employed for balancing out linear trend from main effects and lower order interactions in some factorial designs and in some Latin and Groeco Latin square designs.

Malthus' Theory. According to this theory the human population will expand exponentially while the means of subsistence will increase more slowly, with the consequence that ultimately its size will be kept in check by famine, disease and war. It was given by the English economist T.R. Malthus (1766-1834) The widespread use of birth control methods during this.

century has made much smaller rates of population growth possible ; consequently the theory has been now mainly of historical interest.

Man-hours Function. The same as the labour-time function.

Manifold Classification. If you divide a population into a number of mutually exclusive classes according to some given characteristic and then each class gets divided by reference to some second, third, etc, characteristic, the final grouping is termed as a manifold classification.

Mann-whitney U-Test. Refers to a test of significance which is applicable to two independent samples where measurement is on at least the ordinal level. The test statistic U assesses the probability of a particular pattern of results in relation to the total number of possible patterns. The test regards that the two samples have been independent and that sampling has been random. The hypothesis being tested is that there is no difference between the populations from which the samples were drawn.

Mantel-haenzel Test. It is another name for log-rank test.

Marcinkiewicz's Theorem. According to this theorem (1938) if a characteristic function ϕ is having the form exp $P(t)$ where P is a polynomial then P is a quadratic polynomial and ϕ is the characteristic function of a Normal distribution.

Marginal Category. Refers to one of the frequency classes of a marginal classification.

Marginal Classification. In a bivariate frequency table it is customary to show, as row and column totals, the univariate frequencies of the two variates separately. This is sometimes termed as marginal classification.

Marginal Efficiency of Capital. The internal rate of return.

Marginal Distribution. Refers to the distribution of row totals or column totals in two-way or in multiway tables.

For example in marginals the distributions indicated in the marginals row and column

2, 10, 26, 41, 26, 12, 2 ; and

1, 9, 25, 40, 31, 10, 0

have been marginal distributions of frequencies.

Marginal Probability. In a two-way table, showing the probabilities of outcomes of two events, the row and column totals provide the separate probabilities of outcomes of each of the events.

Marginal Revenue. Refers to the change in gross revenue due to a change of one unit in the rate of sales resulting from a change in price. Its formula is as follows :

$$M = p\left(1 - \frac{1}{e}\right)$$

where M is the marginal revenue, p the price, and e the price elasticity of demand. The price elasticity of demand is an econometric concept of considerable practical significance.

Marginals. These are the total univariate frequencies of two variates in a bivariate frequency table, shown separately in the marginal row and marginal column of the table. The term 'marginal' in this sense means 'at the margin' or 'at the end of the row or column' and must not be confused with the particular sense of 'differential', which the word has in economic analysis.

Markov Chain. A series of discontinuous states at intervals of time $(t_1, t_2, \ldots t_p)$ such that the relative size of constituent classes or groups in a population at any given time is determined by stochastic probabilities inherent in the previous state, and not in any way affected by the earlier state(s) of the system.

Markov Estimate. Refers to an estimate of a parameter derived from an estimator given by the so-called Markov or Gauss-Markov theorem.

Markov Inequality. It a variate x has been non-negative and has mean equal to a then for $t > 0$ the Markov inequality states that

$$\Pr\{x \geqslant t\} < \frac{a}{t}.$$

Some writers owe this inequality to Tchebychev.

Markov Process. Refers to a stochastic process such that the conditional probability distribution for the state at any future instant, given the present state, has been unaffected by any additional knowledge of the past history of the system.

Marshall-edgeworth-bowley Index. It is an index number formula which was proposed by Marshall, Edgeworth and Bowley as an alternative to the standard formulae of Laspeyres and Paasche. It may be written, in terms of prices and quantities, as :

$$I_{on}=\frac{\Sigma\{p_n(q_o+q_n)\}}{\Sigma\{p_o(q_o+q_n)\}}$$

and where the suffix *o* refers to the base period and *n* to the period to which the index relates. The index is a compromise with no general bias.

Marshall-Olkin Distribution. Refers to a multivariate exponential distribution which was introduced by Marshall & Olkin (1967). The distribution is having survivor function.

$$\exp[-\sum_{i=1}^{n} \lambda_i x_i - \sum_{i<1} y_{ij} \max(x_i, x_j)$$
$$-\sum_{i<j<k} y_{ijk} \max(x_i, x_j, x_k) - \ldots$$
$$-\lambda_{12\cdots n} \max(x_1, x_2, \ldots x_n)].$$

Martingale. Originally, this term refers to a process known to gamblers under which the loser at a fair game doubled his stakes for the next, and so on at each loss, the paradox being that in the long run heappeared certain to win sooner or later and at that point would have a net gain.

More recently the term has been assigned a precise significance in the theory of stochastic processes.

Master Sample. Refers to a sample which is drawn from a population for use on a number of future occasions, so as to avoid *ad hoc* sampling on each occasion. Sometimes the Master Sample has been large and subsequent inquiries have been based on a sub-sample from it.

Matched Groups. Refer to groups which are matched for similarity before an experiment is carried out on one of them, the other being used as the control group.

Matched Samples. If the elements in a pair (or set) of samples are such that each element in one sample corresponds to a particular element in the other on some characteristic other than the variable being investigated, the samples are said to be matched.

Matching. If we compare the two sequences of a finite number of characteristics $A_1, A_2, \ldots, A_k$ the *j*th of one sequence against the *j*th of the other, a comparison in which both members exhibit the same characteristic is termed as a match. The number of matches in an observed pair of sequences gives a test of various hypotheses concerning the system which generated them.

Matching Distribution. Suppose two sets having n objects each have been numbered 1, 2,...,n and each has been arranged in a random order so as to form n pairs. Then the number of pairs on which the two numbers have been the same has the matching distribution.

Matrix Sampling. The process of taking a bi-sample from an R×C population matrix whose elements are $\| x_{ij} \|$ has two samples, r of rows and c of columns, and forming a matrix whose elements have been those at the intersections of the selected rows and columns.

Maverick. A term used in industrial statistics to denote an observation lying so far outside the usual range that it has been suspected of not belonging to the population under inquiry.

Maximum F-ratio. In testing the homogeneity of a set of variances, this ratio refers to the ratio of the largest to the smallest.

Maximum Likelihood Estimators. It is a method of estimating the parameters of a population model from the sample data through the likelihood function.

Maximum Likelihood Method. It is a method of estimating a parameter or parameters of a population by that value (or values) which maximises (or maximise) the likelihood of a sample.

Maxwell Distribution. Refers to a fortunately rate expression for a chi-squared distribution having three degrees of freedom. It is based presumably on Clark Maxwell's discussion of the energy of particles moving at random in three dimensions.

Maxwell-Baltzmann Statistic. See Bose-Einstein statistics.

Mean. The most commonly used average or measure or central tendency or measure of location. The mean is computed by adding together the observations in a series and dividing by the number of observations in the series. In statistical symbols, the mean of sample is indicated by the notation $\bar{x}$, and the parameter or population mean is indicated by the letter μ. Thus

$$\bar{x} \; \frac{1}{n} \Sigma x_i \text{ and } \mu = \frac{1}{N} \Sigma X_i$$

where x_i and X_i are the sample and population observations respectively and where n and N are the sample and population sizes respectively.

Mean Absolute Deviation. Refers to the mean of the absolute values of the differences between the values of a variable and the mean of its distribution.

Mean Absolute Error. Refers to an alternative but much less preferable name for the mean deviation.

Mean Deviation. A measure of the spread or dispersion of a sample of population. Deviations of individual observations from a reference point, usually the arithmetic mean, are calculated. Negative signs are ignored so that only magnitudes of deviations are recorded and not their directions. These deviations are then averaged using an arithmetic mean. Thus if $x_1, x_2, \ldots x_n$ are observations and m is their mean (or reference point)

$$MD = \frac{1}{n} \sum_{i=1}^{n} | x_i - m |$$

For grouped data where a frequency f_i is associated with x_i

$$MD = \frac{1}{n} \sum_{i=1}^{k} f_i | x_i - m |$$

where $N = \sum_{i=1}^{k} f_j.$

Mean Deviation, Coefficient Of. Mean deviation expressed as a proportion of the mean. The coefficient is useful for comparing the mean deviations of distributions with dissimilar means.

Mean Difference. Refers to the average of the absolute differences of all possible pairs of values of a variate.

Mean Probit Difference. Refers to a measure of the difference between two series of observations giving parallel probit regression lines.

Mean Range. Refers to the arithmetic mean of the ranges of a set of samples of the same size. The mean range in repeated sampling can be used as an estimator for the population standard deviation.

Mean Range Method (of Estimating Standard Deviation). It is a rapid method of approximating the standard deviation of a population from the ranges of samples. It involves ; (1) the calculation of the mean range of the samples ; (2) the multiplication of the mean range by a factor whose value is determined by sample size. These factors may be obtained from a statistical table, an extract of which is given below.

Mean Semi-Squared Difference. Refers to the statistic which is used in the study of serial variation.

Mean Square. Refers to the arithmetic mean of the squares of differences between a set of given observed values of a variate and a particular value of the variate. In a restricted sense the variance of a set of values has been a mean square, with particular reference to the mean of the observed values of the variate. The term mean square has been more general and may be understood as the mean of squares of differences between observed values and any fixed value.

The term is having a slightly different meaning in analysis of variance. In compiling an analysis of variance table, the sums of squares for the variables are determined. The sums of squares are then divided by the appropriate number of degrees of freedom. Hence, in analysis of variance, mean square is

$$\frac{\text{Sum of squares}}{\text{Relevant degrees of freedom}}$$

Mean Square Deviation/Mean Square Error. Both these terms have been strictly understood in much the same sense as mean square, *i.e.*, the average of the squares of differences between each of a set of observed values of a variate and a given value. If the given value used as a term of reference has been the mean, then all three terms are identical with variance.

The mean square error has been particularly relevant when dealing with biased estimators. Then

$$\text{Mean square error} = \text{variance} + (\text{bias})^2$$

Mean Square Successive Difference. An estimate of the population variance may be based upon the first difference of a series of independent observations, $x_1, x_2, \ldots, x_n$ by the formula

$$\delta^2 = \frac{1}{n-1} \sum_{i=1}^{n-1} \{x_{i+1} - x_i\}^2,$$

δ^2 is termed as the mean square successive difference. Its mean value in random samples from a Normally distributed population has been $2\sigma^2$, whence $\frac{1}{2}\sigma^2$, whence $\frac{1}{2}\delta^2$ affords an unbiassed estimator of σ^2 in the absence of serial correlation in the

In this connection it gets related to von Neumann's ratio: alternative forms using second order differences developed by Kamat & Satde (1962).

Mean Successive Difference. In a time series, this term refers to arithmetic mean of the difference of successive values, *i.e.*, for a series $x_1, x_2, \ldots, x_n$ is

$$d = \frac{1}{n-1} \sum_{i=1}^{n-1} | x_i - x_{i+1} | .$$

Mean Values. Mean values have been a general class of functions of distributions of which moments constitute a special case. If a variate is having a distribution function $F(x)$ and $t(x)$ be some function defined within the range of the distribution, the mean value, or mathematical expectation, may be defined as

$$\xi[t(x)] = \int_{-\infty}^{\infty} t(x) dF(x),$$

subject, of course, to existence. The definition could readily be generalised to n-dimensional variation. The various moments of a distribution have been derived by substituting the appropriate function for $t(x)$.

Measures, Measurement. Both terms may be used to mean a standard common to a number of objects, by which they may be compared (*e.g.*, a centimetre in measuring distance). The term measurement has more extensive uses, which include:

1. systems of measurement (*e.g.*, pound or dollar for measuring money, or centigrade and absolute scales for measuring temperature); and
2. levels of measurement (*e.g.*, cardinal, ordinal, etc.).

Measures of Central Tendency. Typical or central values of a variable. Commonly used measures of central tendency have been mean (or average), mode, median and midrange.

Measure of Location. A quality which is able to purport to locate a distribution, or a set of sample values derived therefrom, by means of a value which is, in some sense, central or typical, *e.g.*, the arithmetic means, the median or the model.

Median. The median may be defined as that value of the variate which divides the total frequency into two halves. As a partition value it could be defined for a continuous frequency distribution by the equation

$$\int_{-\infty}^{M} f(x)dx = \int_{M}^{\infty} f(x)dx = \tfrac{1}{2},$$

M denotes the median value. For a discontinuous variate ambiguity may arise which can only be removed by some convention. For a total frequency of 2N+1 items the median has been the variate value of the (N+1) th item : for 2N items it has been customary to take the average of the Nth and (N+1)th item.

Median Class. Refers to the class in a frequency distribution that contains the median value for a data set.

Median Direction. Refers to a measnre of location for angular data. If a sample of angular observations is considered as a sample of points on the unit circle, then any one point P of these points which has the properties that (*i*) half of the sample points are on each side of the diameter PQ through P, and (*ii*) the majority of the sample points are nearer to P than to Q, is called a median of the sample. The vector $\vec{OP}$ is termed as a median direction of the sample.

Median F-Statistic. If a sample of $2m+1$ items has been ranked and transformed to a uniform distribution the normalised random variable

$$M_{2m+1} = 2\sqrt{(2m+3)}\{F(x'_{m+1}) - \tfrac{1}{2}\}$$

converges in distribution to the standardised Normal variable.

Median Lethal Dose. Refers to a particular name for the median effective dose when the response has been death. It is often written LD_{50}.

Median Line. Refers to any line in a plane which divides a even set of points into two equally numerous sets.

Median Regression Curve. Refers to the type of regression line or curve which is derived from the Mood-Brown procedure.

Median Test. A rank-order test which was proposed by Westenberg (1948) and Mood 19:0). It rejects the hypothesis of identity of two populations, in the one sided case. When there are too few observations from one sample larger or smaller than the median of the combined sample.

Median test is a non-parametric test of similarity between two populations which relates the median of [the two samples to the median of the combined sample.

Median Unbiassed Confidence Interval. It is a $(1-\alpha)$ confidence interval which is bounded by two confidence limits θ' and θ'' at $(1-\alpha)/2$.

Metameter. Refers to a transformed value of a dose or a response, *e.g.*, logarithm or probit which is obtained by using at transformation equation that is independent of all parameters. It has been adopted mainly to simplify the analysis or the expression of the dose response relationship.

Mid-mean. Refers to the arithmetic mean of the sample observations which remain after trimming the smallest and the largest 25 per cent of the original sample. If the original sample size has been not a multiple of 4, the two extreme remaining observations after trimming have been appropriately weighted.

Midrange. A measure of location or central tendency. It is the arithmetic mean of the largest and smallest observations in the data. Thus, if x_1, x_2, ... x_n are a set of numbers arranged in order of magnitude, then the midranges is

$$\frac{x_1+x_n}{2}$$

It is half way between the largest and the smallest of the numbers.

Minimal Sufficient Statistics. A vector of statistics is said to be minimal sufficient if it is having a minimal nnmber of components, in which case it is a function of all other sufficient vectors for the parameters in question.

Minimax Estimation. Refers to the estimation of parameters by the application of the minimax principle to a risk function. In particular it is possible to show that a Bayes' estimator which has a constant risk function has been also a minimax estimator.

Minimax Strategy. If a strategy gets selected from a group of admissible strategies as being the one which, on a basis of the expected loss, is having the smallest maximum loss this strategy would be a minimax strategy.

Minimum Chi-Squared. Refers to a method of estimation which is based upon the χ^2 goodness of fit statistic. The method determines values of the parameters so as to minimise χ^2 calculated from observed frequencies and 'expected' frequencies expressed in terms of the parameters.

Minimum Discrimination Information Statistic. This statistic has been based upon the principle of information in a sample and can be regarded as the 'divergence' of the alternative hypothesis from the null hypothesis. The distribution of this statistic is involving central and non-central χ^2 and it is having additive properties. For a wide class of problems concerning contingency table, this statistic takes the form of $-2 \ln \lambda$ where λ refers to a likelihood ratio.

Minimum Probability. Refers to probability of selling at least an additional unit that must exist to justify stocking the unit.

Minimum Variance. When applied to estimators, this term is used to represent the property of possessing the least variance among the members of a defined class. A minimum variance estimator exists only where there is existing a sufficient estimator.

Minimum Variance Linear Unbiassed Estimator. In the case of the linear hypothesis the method of least squares gives estimators that have been unbiassed, linear in the observations, and with minimum variance.

MINQUE. Rao (1970) introduced a method called MINQUE (minimum norm quadratic unbiassed estimation) which was used for estimating the variances σ_i^2 for the Gauss-Markov linear model

$$Y = X\beta + e$$

where Y denotes a vector of n observations, X an $n \times m$ known matrix, β a vector of m unknown parameters and e an error vector with a dispersion matrix of diagonal form.

Missing Plot/Missing Value. In an experiment or a survey, some planned observations may be missing because of circumstances beyond the control of the researcher. These cases may be given a special code and special techniques are available for analysing data where some of the values are missing, but these are outside the scope of this book. The term 'plot' arises from the fact that most such early work was done in the area of agricultural experimentation where a missing plot was one for which, for some reason, no observation was available.

The use of plot as in plotting is completely different.

Missing Plot Technique. It is the name assigned by Allen & Wishart (1930) and Yates (1933) to the process of analysing material which had been designed to conform to an experimental pattern but from which certain values have been missing through circumstances beyond the control of the experimenter. The use of the word 'plot' arose from the agricultural background of the original investigations noted above but methods have been applicable to missing values generally.

Mitscherlich Equation. It is possible to estimate the mounts of an element in the soil using Mitscherlich's equation

$$Y=A(1-bR^x)$$

where Y denotes yield and x the amount of fertiliser applied.

Mixed Autoregressive-Moving Average (ARMA) Process. Refers to a stochastic process $\{u_t\}$ in discrete time which is driven, by independent and identically distributed random disturbances $\{e_t\}$ with expectation zero, in accordance to the model.

$$(1-\phi_1B-\phi_2B^2-\ldots-\phi_pB^p)(u_t-\mu)$$
$$=(-1\theta_1B-\theta_2B^2-\ldots-\theta_qB^q)\epsilon_t$$

i.e., $$\phi(B)\bar{u}_t=\theta(B)\epsilon_t$$

where B denotes the unit time backward shift operator,. The non-negative integers (p, q) are jointly termed as the order of the ARMA process.

Mixed Autoregresive-Regressive Systems. Refers to a system of equations which could be able to represent a particular econometric model might be an autoregressive set of y's regressed upon fixed x's :

$$\sum_{j=1}^{k} a_j y_{t-j}=\sum_{l=1}^{q} \beta_{lt}x_t+\epsilon_t.$$

Mixed Distribution. A term which hae been frequently used in modern statistical literature for the distribution which gets resulted from the parameter of a given distribution being itself a random variable.

Mixed Model. This term has been used in at least four slightly different senses ; a model is said to be 'mixed' if

(*i*) its equations are having both determinate and stochastic elements ;

or

(*ii*) its equations are having both difference and differential terms ;

or

(*iii*) it is having endogenous as well as exogenous elements ;

or

(*iv*) in an analysis of variance context for a two way layout, the rows correspond to a model I factor, *i.e.*, fixed effects, and the columns correspond to a model II factor, *i.e.*, random effects. This version is some-times termed as model III but this is best avoided (see Plackett, 1960).

Mixed Sampling. A sampling plan is said to be mixed when it consists of a number to different sampling plans or techuiques. For example, it may be decided to overcome the economic disadvantages of simple random sampling by using area sampling or cluster sampling in one context and systematic sampling in another.

Mixture of Distributions. Refer to a process of getting a weighted average of a group of distribution functions which is a new distribution function. It has to be distinguished from the compound distributions.

Modal Group. Occasionally a set of numbers or observations are arranged and summarized as a grouped frequency distribution. The modal group is one with greater frequency for its breadth than either adjacent group. A frequency distribution can have more than one modal group. Modal groups can be defined for non-numeric nominal or ordinal scale frequency distributions.

Mode. The value in a set of numbers which occurs more frequently than neighbouring possible values. There may be more than one mode. For example, the mode of ten numbers 1, 1, 2, 3, 3, 3, 7, 8, 9, 9 is 3 ; the fifteen numbers 2, 2, 2, 2, 3, 3, 5, 5, 5, 5, 6, 6, 7, 8, 9 have two modes, 2 and 5. The latter set of numbers is said to be bimodal. The normal distribution is unimodal. It has one mode, which is coincident with its mean and median. Asymmetric, or skewed distributions, also have single modes which do not coincide with their means and medians. In contrast, some frequency distributions are bimodal and multimodal, usually because the frequency distributions are themselves composites of two or more frequency distributions or different categories, which have differing modes.

A rectangular distribution is having no mode.

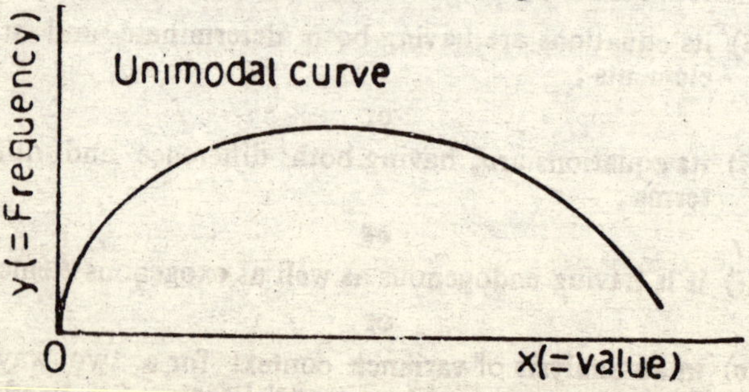

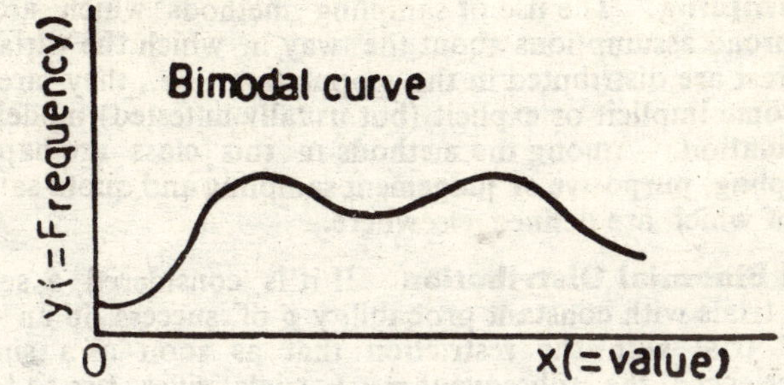

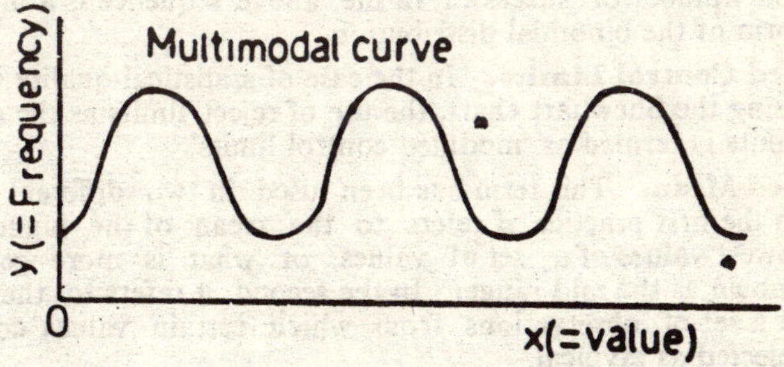

Although the idea of the most frequently encountered or fashionable value of the variate for the mode is probably very old, it was not generally used in statistics until popularised by K. Pearson (1894). The concept has been essentially of use only for continuous distributions, although it can get extended to the discontinuous case. More formally, if $f(x)$ has been a frequency function, a mode is a value of x for which

$$\frac{df(x)}{dx}=0, \quad \frac{d^2f(x)}{dx^2}<0.$$

There may thus be more than one mode of a distribution, though the practical occurrence of multimodality has been comparatively rare.

Model. In statistics a model is generally expressed in terms of symbole (*i.e.*, in mathematical form), usually as a set of equations. However, it is possible also to use diagrammatic models but for the purposes of estimation symbolic representation is usually necessary.

Modeling Techniques. Methods for deciding which variables to include in a regression model and the different ways in which they can be included.

Model Sampling. The use of sampling methods which are based on broad assumptions about the way in which the variables of interest are distributed in the population, *i.e.*, they are based on some implicit or explicit (but usually untested) model of the population. Among the methods in this class are haphazard sampling, purposive or judgement sampling and quota sampling, all of which are defined elsewhere.

Modified Binomial Distribution. If it is considered a sequence of n trials with constant probability p of success in an individual trial with the restriction that as soon as a success is experienced, the subsequent $m-1$ trials gives rise to failures. The number of successes in the above sequence is a modified form of the binomial distribution.

Modified Control Limits. In the case of statistical quality control using the Shewhart ehart, the use of reject limits as the control limits is termed as 'modified control limits'.

Modified Mean. This term has been used in two different senses. In the first practice, if refers to the mean of the highest and lowest values of a set of values, or what is more generally known as the mid-range. In the second, it refers to the mean of a set of observations from which certain values could be rejected as atypical.

Modulus. A number expressed in terms of its value without regard to its sign. The average deviation is derived by using the moduli of mean deviations. In some contexts it means a mathematical constant.

Moments. A term that is tending to fall into disuse as superfluous ; still seen occasionally in current literature in such phrases as the *method of moments* and *moments about the mean*. There are three moments :

1. The average deviation
2. The standard deviation
3. A measure of skewness which is sometimes called the third moment.

Moment-Generating Function. The moment-generating function of the distribution of the random variable X is given as follows :

$$E(e^{xt}) = \Sigma_i e^{x_i t} p(x_i) \text{ (for a discrete variable)}$$
$$= \int_{\text{all } x} e^{xt} f(x)\, dx \text{ (for a continuous variable).}$$

where $p(x_i)$ denotes the probability distribution and $f(x)$ the probability density function, Symbol : $\mathbf{M}_x(t)$.

Moment Ratio. Refers to a ratio in which the numerator and the denominator have been moments or simple functions of moments.

Moments, Method of. Refers to a method of curve fitting which proceeds by identifying the lower moments of the observed data with those of the particular curve form being fitted.

Monte Carlo Method. A simulation technique originating in operational research designed for solving queueing problems, *i.e.*, for finding the organizational method that would probably minimize the waiting time of men and things in such situations as ships waiting for a vacant berth in a port or an aeroplane waiting for permission to land. The technique employs mathematics and probability theory applied to a number of proposed alternative organizational methods.

Monthly Average. By analogy with annual averages and moving averages generally this term is used to refer to the average of values of a time series occuring within a month, the resulting figure being representative of that particular month.

Most Efficient Estimator. Refers to an unbiassed estimator whose sampling variance has not been greater than that of any other unbiassed estimator has been called a most efficient estimator, the qualification 'unbiassed' being understood. For biassed estimators the same expression has been sometimes used to denote an estimator for which the mean square error is minimal.

Most Powerful Critical Region. Refers to the critical region which is having the highest power in testing a hypothesis.

Most Powerful Rank Test. Refers to a sample rank order test which has been most powerful under some alternative hypothesis ; for example, Normal distributions differing only in the mean.

Most Stringent Test. A test of a statistical hypothesis (H_0) has been said to be most stringent if it minimises the maximum difference by which the test falls short, with respect to a particular class of alternative hypotheses (H_1), of the power that could be attained with respect of these alternatives. A uniformly most powerful test has been necessarily most stringent.

Moving Annual Total. It is a series which is derived from an observed time series in which each term consists of the current observation and those immediately preceding it for the period of a year, *e.g.*, the moving annual total of a monthly series would consist of the sums of twelve consecutive monthly values.

Moving Average. Refers to a form of average which has been adjusted to allow for seasonal components or cyclic components. When a variable, like the cost of a sack of potatoes, or the number of unemployed, is graphed against time, there are likely to be considerable seasonal or cyclic components in the variation. These may make it difficult to see the underlying trend. These components can be elimidated by taking a suitable moving average.

Moving Average Disturbance. In an equation expressing a relationship between variates it has been sometimes convenient to include a final term which serves to summarise the effect of factors not separately specified. If such a term z_t takes the form of a moving average process—say,

$$z_t = a_0\epsilon_t + a_1\epsilon_{t-1},$$

then the equation is considered to have a moving average disturbance.

Moving Average Model. Refers to the representation of a stationary stochastic process in terms of a moving average of infintte length. The observation x_t has been a linear combination of residuals z_t of the form

$$\{x_t\} = \sum_{j=0}^{\infty} b_j z_{j-j}.$$

Moving Average Process. Refers to a special case of the moving summation process. If $\{\epsilon_t\}$ be a random process the process $\{\xi_t\}$ defined by

$$\xi_t = \sum_{t=0}^{k} a_j \epsilon_{t-j}$$

will exist and be stationary if $\{\epsilon_t\}$ is so. It is termed as a moving average process.

Moving Observer Technique. Refers to a method of enumerating a moving population in which the observer himself is having among the population.

Moving Range. This concept is similar to that of moving average. If the total number of measurements on a variable x has been N, by taking absolute differences of successive pairs $|x_t - x_{t-1}|$, we obtain N−1 values which can be recorded as (moving) ranges of successive samples, each of size two. This could be generalised to $| x_{i+k} - x |$, $i=1, 2, \ldots, n-k$ so as to afford a greater degree of smoothing.

Moving Seasonal Variation. Refers to a pattern of seasonal variation which changes with time. It is generally obtained by determining a seasonal pattern for a certain number k of consecutive years and 'moving' the set of k years along the series as for a moving average, so getting a seasonal pattern for each year based on the previous k years.

Moving Weights. In most cases where a moving average may be taken of a time series the weights composing the average have been constants independent of time.

Mth Values. The mth values of a set of n observations have been the mth largest or mth smallest when the values have been arranged in order of magnitude. They have been particular cases of order statistics. Thus, if $m=1$ the mth values have been the two extreme values.

Multi Asset Project. A proposed capital work in which it is planned to provide two or more fixed assets with different book lives, so that separate provision for renewal or amortization of each asset has to be made.

Multicollinearity. In regression analysis, this term refers to a situation in which there exists a linear relation connecting the predicated ('independent') variables have been then indeterminate and their standard errors become infinite.

Multi-equational Model. Refers to a model of a system where the variables have been interconnected by more than one equation. An alternative name has been simultaneous equations model.

Multimodal. Having many modes. Frequency distribution curves may consist of many modes when they are the composites of a number of distribution curves.

Multiphase Sampling. It is sometimes convenient and economical to collect certain items of information on the whole of the units of a sample and other items of information (usually more intensive of detailed) only for a subsample of the units. This is called two-phase sampling. Further phases may be added as required. All such sampling schemes are called multiphase.

Multiple Bar Chart. Refers to a chart which shows two or more characteristics in the form of bars of length proportional to the magnitude of the characteristics.

Multiple Classification. It is an alternative name for manifold classification. Sometimes the expression has been restricted to the case of quantitative variates.

Multiple Curvilinear Correlation. This term is used to describe the relationship between variates which are not having linear regression one on another but are not independent.

Multiple Decision Methods. Refers to the statistical techniques which are used for dealing with problems where the possible outcomes or decisions have been more than two.

Multiple Decision Problem. Refers to the problem of choosing one hypothesis or decision from a set of k mutually exclusive and exhaustive hypotheses or decisions on the basis of some observations on a random variable. The simple problem of testing a statistical hypothesis against a single alternative has been a special case of this, when $k=2$.

Multiple Factor Analysis. Now this term is equivalent to factor analysis. Historically, this term refers to the analysis of psychological material into factors grew from one factor to two factor and then to m-factor complexes and the latter was termed as 'multiple' to distinguish it from the simpler forms. It no longer seems necessary.

Multiple Markov Process. Rafers to a stochastic process in which the transition probabilities have been found to depend on previous values at more than one point. In most contexts the expression has been equivalent to 'autoregressive process'.

Multiple Poisson Distribution. This distribution has been the joint distribution of independent Poisson variates. It may also be considered as an approximation to the multinomial distribution when m has been large and $np_i=\lambda_i$ in the form :

$$e^{-(\lambda_1+\dots\lambda_{r-1})}\frac{\lambda_1^{k1}\lambda_2^{k2}\dots\lambda_{r-1}^{k_r-1}}{k_1!k_2!\dots k_{r-1}!}$$

Multiple Poisson Process. This term is sometimes used for a generalisation of a Poisson process where there have been simultaneous occurrences at the event points.

Multiple Random Starts. Refers to an idea for selection procedure under systematic sampling. The method has been to choose a random sample of s without replacement from the first k elements and subsequently every kth multiple element of those selected.

Multiple Range Test. It is a method of comparing mean values which are arising in analyses of variance by using the range of subsets of means, or of ranges of sets of observations contributing to mean values.

Multiple Recapture Census. Refers to a sequential method of capture/release sampling where the method of capture does not kill or influence the behaviour of the sample unit and where the experimenter is having full control over the sampling and marking.

Multiple Regression. 1. Refers to the statistical process by which several variables have been used to predict another variable.

2. The regression of a dependent variate on more than one 'independent' or 'predicted' variable.

Multiplication Law. A law of chance in which two or more probabilities are multiplied together to give a total probability that each of two or more events will occur.

Multiplication Principle (Probability). This principle states that (1) if two outcomes are independent of each other, the probability of both outcomes occurring is the product of the probability of each of the outcomes and (2) if the two outcomes are such that one outcome is dependent on the other, the probability of both outcomes occurring is the probability of the first event multiplied by that of the second outcome given the first.

Multiplication Rule. Refers to the rule which states that, for two events A and B, the probability p of their both occurring is given by ;

$$p(A \cap B) = p(A \mid B).p(B)$$

or

$$p(A \cap B) = p(B \mid A).p(A)$$

If the events A and B have been independent, this can be simplified to

$$p(A \cap B) = p(A).p(B)$$

Multiple Stratification. If a sample has been stratified according to two or more factors it is considered to be multiply stratified.

Multiplicative Process. This term is synonym of branching process.

Multi-Stage Sampling. Refers to a sample which has been selected by stages, the sampling units at each stage being sub-sampled from the (larger) units chosen at the previous stage. The sampling units pertaining to the first stage are termed as primary or first stage units ; and similarly for second stage units, etc.

Multivariate Analysis. This expression has been used rather loosely to represent the analysis of data which have been multivariate in the sense that each member bears the values of p variates.

The principal techniques of multivariate analysis, beyond those admitting of a straightforward generalisation, *e.g.*, regression, correlation and variance analysis, have been factor and component analysis, classification, discriminatory analysis, canonical correlation and various generalisations of homogeneity tests, as illustrated by the D^2 statistic.

Multivariate Binomial Distribution. Refers to an s-variate generalisation of the binomial distribution. For example, if a member can bear the value of two dichotomised variables having probabilities

		Variable 1	
		Present	Absent
Variable 2	Present	p_{11}	p_{10}
	Absent	p_{01}	p_{00}

the bivariate binomial in samples of n would be given by $(p_{11}+p_{10}+p_{01}+p_{00})^n$.

Multivariate Distribution. Refers to a distribution of more than one variable. If two variables are involved it is called bivariate, if three, trivariate and so on Compare Bivariate distribution.

Multivariate Exponential Distribution. Refers to a distribution of considerable importance in life assessment of systems. The complement of the distribution function has been given by $\Pr\{x_1>_1, X_2>x_2, ..., Xx_n>x_n\}$ which has been equal to

$$\exp[-\sum_{s\in s} \lambda_s \max (xis_i)],$$

where S has been a set of vectors : marginal distributions have been ordinary negative exponential.

Multivariate 'F' Distribution. If there have been two nonsymmetric matrices $m\times p$ and $m\times n$ $(p\leqslant m\leqslant n)$ denoting matrix variates X and Y and all columns have been Normally and independently distributed with covariance Σ, the multivariate F-distribution has been that of $X'(YY')^{-1}X$ provided $\xi(X)=\xi(Y)=0$.

Multivariate Hypergeometric Distribution. Let n balls be drawn at random without replacement from an urn which has N balls of $(s+1)$ different colours, M_i being of the ith colour, $i=0, 1, 2, ..., s$. If X_i denotes the number of balls of the ith colour $i=1, 2, ..., s$ drawn in the n draws then $(X_1, X_2, ..., X_s)$ is having the s-dimensional hypergeometric distribution).

This distribution is also termed as the factorial multinomial distribution.

Multivariate Poisson Distribution. Refers to an s-variate generalisation of the Poisson distribution and the limiting distribution as $n \to \infty$ and $p_i \to 0$ of the multivariate binomial distribution.

Multivariate Processes. Refers to a class of stochastic processes which involves more than one random variable. It is also termed as simultaneous or vector processes and not to be confused with multi-dimensional processes which are concerned with more than one parameter.

Multivariate Quality Control. Refers to control of quality in which each item for inspection must be able to conform to standards in respect of more than one variable, as, for example, the length, breadth and height of a metal block.

Mutually Exclusive. Two events are mutually exclusive if the occurrence of one of the events renders the occurrence of the other event contemporaneously imposlble. Thus it is not possible for a person to throw a single coin and obtain both a head and a tail at the same time, nor is it possible for a candidate both to pass and fail the same examination at the same time.

Mutually Exclusive Classification. The division of a population, sample or set of observations into a set of classes or categories which are mutually cxclusive. The classes may either be attribute categories (*e.g.*, sex, race, occupation, nationality) or variate interval classes (1—10, 11—20, 21—30, 31—40). The classification may either be univariate (as represented in a histogram or other single distribution of mutually exclusive categories); or it may be bivariate (*e.g.*, the $m \times n$ table of mutually exclusive categories required for chi-squared or modified chi-squared tests); or it may be multivariate.

Mutually Exclusive Events. Events that cannot happen together.

Mutual Regression. A national formula of the method of least squares (L7) in which the burden of error would be distributed equally among the several variables. In the regression of Y on

X, for instance, Y carries the burden of error, *i.e.*, Y and not X is made subject to all the variations due to factors other than X in the relationship between the two factors. Mutual regression would take a middle course between the regressions of Y on X and X on Y.

N

Negative Binomial Distribution. Refers to a distribution in which the relative frequencies (probabilities) have been arrayed by a binomial with a negative index. For example, if the variate values have been 0, 1, 2, etc., the frequency at $x=j$ is the coefficient of t^j in the expansion of $(1-pt)^{-n}$ $(1-p)^n$ in powers of t. The distribution is sometimes termed as the Pascal distribution.

Negative Moments. The negative moments of a frequency distribution may be defined as the moments of reciprocal powers of the variate ; for example, $\xi(1/x^k)$ where k denotes the order of the moment Their existence, of course, depends upon convergence and they have been little used. In fact, owing to possible confusion with ordinary moments, which have negative sign, the preferable expression would have been 'moments of negative order.

Negative Multinomial Distribution. If there has been a sequence of independent trials in each of which there can be $s+1$ mutually exclusive outcomes $A_0,\ldots,$ A_s having probabilities $p_0,\ldots,p_s$ $(\Sigma p_i=1)$ and $x_1,\ldots,x_s$ are the occurrences of $A_1,\ldots,$ As before A_0 occurs k times, then $(x_1,\ldots,A_s$ (is having the negative multinomial distribution.

Nested Balanced Incomplete Block Design. Refers to an experiment design having two systems of blocks, the second within the first such that ignoring either system leaves a balanced incomplete design whose blocks have been those of the other system. In the nesting system each block from the first part is having m blocks from the second.

Nested Classification. If one of the factors in a two-way classification has been such that no main effect could be attributed to it, then that factor may be said to be nested within the main classification.

Nested Design. Refers to a class of experimental design in which every level of a given factor is appearing with only a single line of any other factor. Factors which are not nested are called crossed. If every level of one appears with every level of the others, the factors have been said to be completely crossed : if not, they are partly crossed.

Nasted Sampling. A term which is used in two somewhat different senses : (1) as equivalent to multi-stage sampling because the higher-stage units get 'nested' in the lower-stage units ; (2) where the sampling has been such that certain units get imbedded in larger units which form part of the whole sample, *e.g.*, the entry-plots of clusters get 'nested' in this sense.

Net Correlation. Refers to an alternative but absolescent term for partial correlation. The name has been derived from the fact that a partial correlation has been the correlation between two variates of a larger group when the remaining variates have been held constant giving, in a sense a 'net' correlation.

Net Error. Refers to that part of the total or gross error which does not cancel out when cases or observations have been averaged or totalled.

Net Present Value. One of the three principal techniques and yield criteria of discounted cash flow ; it is equal to the present value of the net yield of a fixed asset over its life less the outlay on the asset. In the discounting process an appropriate current rate of interest is used. The lower the rate of interest used for the purpose, the greater is the calculated present value of the yield. The present value of Rs. 1000 a year for 20 years at 15% is Rs. 6259 ; at 10% it is Rs. 8514 ; and at 5%, Rs. 12462. The net yield of the asset should be gross of depreciation, as the NPV technique embodies sinking-fund depreciation provision and also charges interest on capital—both at the rate of interest used in the discounting process.

Net Regression Coefficient. The coefficient of an independent variable in an analysis involving a least one other independent variable ; any coefficient in a multiple regression equation.

Net-Revenue Schedule. A statistical schedule which sets out a range of prices of a brand of product, and the corresponding annual sales in terms of quantity and money, the annual total or variable costs and the profit. It is demonstrated in Table 1.

Network of Samples. Refers to an alternative name for a set of interpenetrating samples.

Neutral Curve (Curva Neutra). Refers to the frequency curve of a distribution having neutral abnormality.

Neyman Allocation. A name given to optimal allocation of a sample when casts are assumed to be equal in each stratum.

Neyman Model. This term was given by Ogawa (1963) and used to denote experiment designs which involve situations containing technical errors in the sense that variations in replicate observations have been solely due to experimental techniques.

Neyman—Pearson Lemma. The Neyman—Pearson lemma provides a methodology for finding the best test among those with the same probability of type I error, *i.e.*, the test with the highest power (the lowest probability of type II error).

The test is based on the likelihood ratio and an example is given in the entry likelihood ratio test.

Node. Point at which a chance event takes place on a decision tree.

Nominal Scale. The simplest form of measurement has been on a nominal scale. Here, individuals or elements get allocated to categories of the characteristic being measured where no relationship is assumed to exits between the categories.

Nomogram. A form of multiple line diagram in which the scales for different functions of a variable are drawn alongside each other and corresponding values are in line with each other. Nomograms facilitate the indentification of differences between scales, but are not as accurate as tables

Non Causa. False cause ; a term used in logic for the fallacy of assuming an unproved cause.

Noncentrality Parameter. Refers to a parameter of the distribution of a noncentral test statistic that gets vanished in the central case.

Non-Central Beta Distribution. Refers to the distribution of the ratio

$$\frac{\chi'^2 v_1}{\chi^2_{v1}+\chi^2_{v2}}$$

where χ^2_{v1} has been the noncentral χ^2. If both χ^2 have been con-central the resultant distribution has been the doubly non-central beta distribution.

Non-Central Confidence Interval. Refers to a confidence interval which has been not central.

Non-Central χ^2 Distribution. The distribution of the sum of squares of independent Normal get variated with unit variance but not with zero mean. The distribution was given by R. A.

Fisher in 1928 as a special case of the distribution of the multiple correlation coefficient.

Non-Coverage. Failure to include units or sections of the survey population in the operational sampling frame.

Non-Determination, Coefficient of (k^2). Refers to a measure of alienation, which is the square of the coefficient of alienation, and related to the product-moment correlation coefficient (ρ).

$$k^2=1-\rho^2$$

The coefficient meaures the proximity of relationship between two variates to zero-correlation.

Non-Linear Correlation. This term has been meant to relate to the correlation between variates where the regression has been not linear. It is thus a misnomer; correlation, being a pure number, cannot be non-linear.

Non-Linear Interpolation. See interpolation.

Non-Linear Regression. The expression for the regression line may be non-linear in the coefficients; or in the explanatory variable, *e.g.*, a polynomial relationship.

Non-Normal Population. Refers to a population for which the frequency distribution has been not the Normal distribution. The term does not imply abnormal in the sense of unusual.

Non-Null Hypothesis. In general it refers to a hypothesis alternative to the one under test; the null hypothesis. In some context, however, it has been given the meaning of a hypothesis under test where the effect has been not equal to zero.

Non-Orthogonal Data. Data have been said to be non-orthogonal if they give rise to estimates of various effects which are not independent of one another or, perhaps or other features such as block differences which are a nuisance in the analysis.

Non-Parametric. See distribution-free method.

Non-parametric Statistics. That branch of statistics comprising techniques of statistical analysis that do not entail assumptions about the exact form of the distribution of the population. The use of nonparametric statistical tests does not require that the sample being analyzed be from a population with a normal distribution or with any other specified distribution (that is, the exact shape of the distribution of the population need not be known.

Non-Parametric Tests of Significance or Distribution-free Tests of Significance. Tests of significance which make no assumptions concerning the parent distribution. All tests involving the ranks of data have been non-parametric.

Non-Parametric Tolerance Limits. Tolerance limits which are not dependent on the parameters of the parent population from which a sample is drawn.

Non-Random Sample. A sample which is selected by a non-random method. For example, a scheme whereby units have been selected purposively would provide a non-random sample. Again, a sample obtained by taking members at fixed intervals on a list has a non-random sample unless the list get arranged in a random order.

Non-Ragular Estimator. See regular estimator.

Non-Response. The failure to get the required information from members of the selected sample for any reason. In order to measure the rate of non-response it becomes necessary to keep careful and complete records of the selection of individuals and the outcome of the fieldwork. The non-response rate may be defined as the proportion of designated individuals for whom the survey information was not obtained.

Non-Sampling Errors. Refer to errors in estimates which have been not due to the fact that the information has been based on a sample from the population. Thus non-sampling errors are important even when information has been obtained for the whole population, as in the case of the Census of Population.

Non-Singular Distribution. Refers to a multivariate distribution in, say, p variates which cannot, by a linear transformation of the variates, be converted into a distribution with fewer than p variates. A distribution has been non-singular if and if the dispersion matrix or the correlation matrix has been of rank p.

Norm. Refers to an average level of achievement, performance or behaviour.

Normal Curve, Normal Distribution. Refers to the most commonly used type of continuous frequency distribution (or distribution curve). It coincides with the distribution of many observations of natural phenomena (such as height, weight and ability), and has been also the limiting form of the binomial distribution where $p=q$. The curve is bell-shaped and symmertical.

The curve and associated distribution have been often called Gaussian and Laplacean, because of their association with Gauss and Laplance.

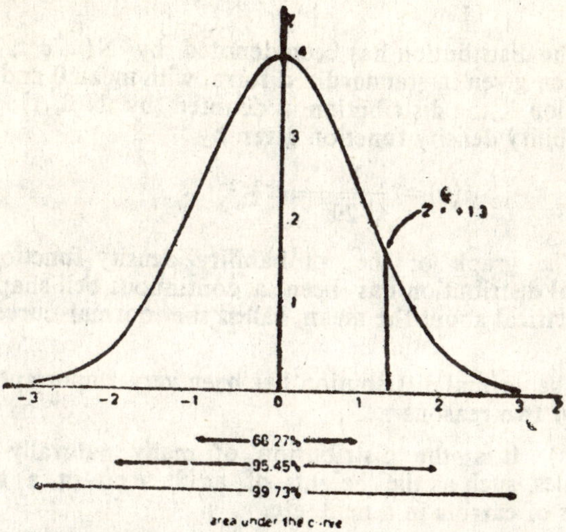

Normal Curve of Error. A symmetrical frequency-distribution curve ; variously called the ***normal probability curve***, the ***normal frequency-distribution curve*** and, in its context, simply the ***normal curve***.

Normal Deviate. Refers to the value of a deviate of the Normal distribution.

Nr mal Distribution. Refers to a continuous distribution of a random variable with its mean, median and mode equal. Symbol : $N(\mu, \sigma^2)$.

The probability density function $\phi(x)$ of the normal distribution, with mean μ and standard deviation σ, has been given by :

$$\phi(x) = \left(\frac{1}{\sqrt{2\pi\sigma}}\right) \exp\left\{-\frac{(x-\mu)^2}{2\sigma^2}\right\}$$

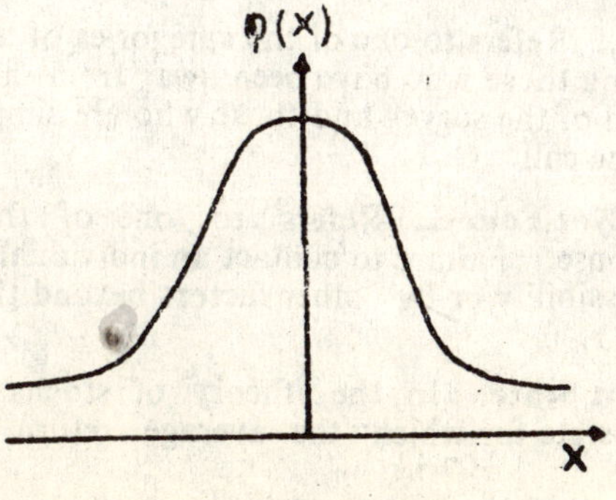

Fig. 1. Normal curve.

The distribution has been denoted by $N(\mu, \sigma^2)$. When it has been given in standardized form, with mean 0 and standard deviation 1, the distribution is denoted by $N(0, 1)$, and has probability density function given by :

$$\phi(x) = \frac{1}{\sqrt{2\pi}} e^{-\frac{1}{2}x^2}$$

The graph of the probability density function of the normal distribution has been a continuous bell-shaped curve, symmetrical about the mean, called the normal curve or *error curve.*

The normal distribution has been very important in statistics for two reasons :

(*a*) It is the distribution of many naturally-occurring variables, such as the heights of adult men in a town, the masses of carrots in a field, etc.

(*b*) The distribution of the means of samples drawn from most parent populations is normal or approximately so, when the samples are sufficiently large.

Normal Equations. A set of simultaneous equations derived in the application of the least squares method, for example in multiple regression analysis. They are used to estimate the parameters of the model.

Normalisation of Frequency Function. Refers to the transformation of a variate so that its frequency function becomes Normal, or approximately so.

Not Applicable/Does not Apply. A classification which is used to distinguish the situation where a question is not applicable to a particular respondent from a refusal to answer the question.

Not-at-homes. Refers to one of the categories of non-response. It is including those who have been away from home throughout the period of the survey and those who are simply out at the time of the call.

Not Reached/Not Found. Refers to one of the categories of non-response. Failure to contact an individual may get caused by inaccessibility or by other factors beyond the control of the interviewer.

Null Recurrent State. In the theory of stochastic processes, a recurrent state for which the average return time has been infinite.

Nyquist Frequency. In connection with data consisting of observations recorded at equal distances of time, the frequency of a sinusoidal term whose period has been twice the time interval between successive observations.

Nyquist Interval. For time series having a series of harmonic components with frequencies in a limited range, the Nyquist interval has been such that the highest admissible frequency has a period of two intervals.

O

Oblique Factor. In educational or psychological testing it refers to a factor which gets correlated with one, or more, other factors has been said to be oblique. In one geometrical representation on the situation the vectors which represent them have been no longer orthogonal one to another. The use of oblique factors gets confined almost entirely to psychology.

Observable Variable. Refers to a mathematical or stochastic variable, the values of which could be directly observed, as distinct from unobservable variables which enter into structural equations but have not been directly observable.

Observation. (1) The observed value of a variate for an element is termed as an observation ; and (2) observation is the process of getting information directly, for example, by observing behaviour—as distinct from obtaining information from reports or records—by asking people about their behaviour, for example.

Observational Error. Refers to an error of observation but sometimes occurs as meaning a response error.

Occupancy Problems. Refers to the general class of problems in probability which are dealing with the random distribution of *r* objects over *n* cells with particular reference to the numbers of objects falling in particular cells.

Octiles. Refer to the set of seven variate values which would divide the total frequency into eight equal parts.

Odds. Refer to the ratio of the probability of an event occurring to that of its not occurring.

Odds Ratio. Refers to an alternative term for relative risk.

A technique for specifying the nature of an office/plant for maximum system performance.

Ogive. Refers to a continuous form of the cumulative frequency curve, which considers a very large number of observations. If the cumulative frequencies have been plotted against the ordinate axis and the variate values have been plotted against the abscissa, any unimodal cumulative frequency distribution will have an S-shaped appearance, taking the form illustrated see page 656.

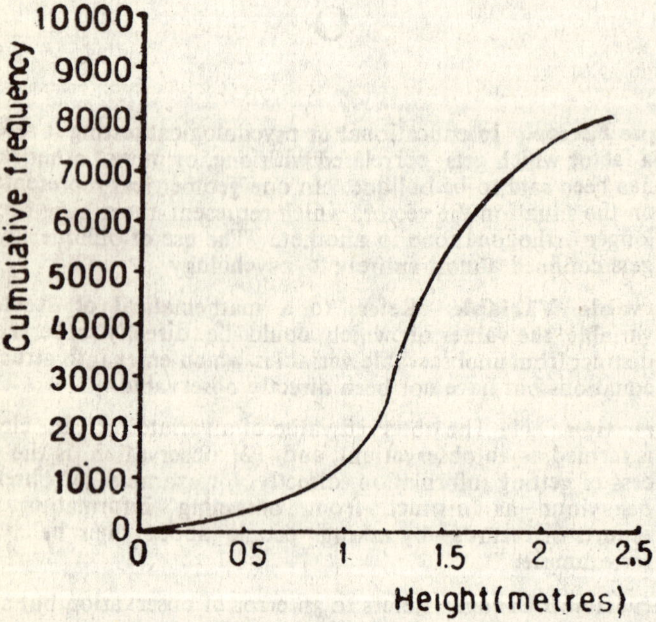

An ogive which shows the measurements of people under and including a given height.

One-Sample Runs Test. A nonparametric method for determining the randomness with which sampled items have been selected.

One-Way Analysis of Variance. See analysis of variance, kruskal-wallis one-way analysis of variance.

Open-Ended Class. A class that allows either the upper or lower end of a quantitative classification scheme to be limitless.

Open Sequential Scheme. Refers to a sequential sampling scheme which does not but an ultimate limit to the size of the sample.

Open Ended Classes. When in a frequency distribution, the initial class interval has been indeterminate at its beginning and/or the final class interval has been indeterminate at its end, the distribution has been said to possess 'open ended' classes.

Open Question, Open-ended Question. In an open question it has been the respondent who decides the form, length and content of the answer. The interviewer must record as much of the answer as possible.

Operating Characteristic Curve. In quality control, this curve refres to the graph of the probability of accepting a batch from which a sample has been drawn for inspection, against the true proportion of defectives for the batch.

Operational Research. Mathematics and economic theory applied to the field to business management and control, the object being to provide data for policy and executive decision making. Includes linear programming, break-even analysis and profit planning. Some of the mathematical techniques were in management use in the thirties, long before they were blessed with a name. Operational research involves the analysis of problems in business and industry using mathematical and statistical techniques.

Opinion Survey. Refers to a sample survey which aims at ascertaining or elucidating opinions possessed by the members of a given human population with regard to certain topics.

Optimal Asymptotic Test. An alternative name assigned to a locally asymptotically most powerful test.

Optimality of Design. In the design of an experiment of regression using the linear model, it has been possible to adopt certain desirable criteria which have been optimal in some sense. One natural criterion for parameter (s), a given confidence coefficient and residual sum of squares arising from a particular design, has been to make the content of the confidence ellipsoid as well as possible. This criterion is designated D optimality. If the primary interest is lying in $\xi(\overline{y})$ over some region then, if this region is coinciding with the design space, the criterion is designated G optimality. If, however, the region has been the unit sphere, then the criterion is known as E-optimality.

Optimum Allocation. In general, this term refers to the allocation of numbers of sample units to various strata so as to maximise some desirable quantity like precision for fixed cost.

Optimum Linear Predictor. Refers to a predictor for future observations of a stochastic process which gets constrained to be a linear combination of past observations and which minimises the mean square error of prediction.

Optimum Statistic. Refers to an expression which has been usually synonymous with best estimator. If it refers to a statistic used for testing hypotheses it is usually termed as an optimum test statistic.

Optimum Stratification. Refers to a system of stratification in sampling which optimises some given criteria. Generally the criteria relate to a set of estimated implies and needs the minimisation of their generalised variance.

Optimum Test. Refers to a test which can be shown to have a certain desirable characteristic or group of characteristics to a greater degree than any other test of the same class.

Optimum Value. Refers to that value which has the effect of maximizing another value, *e.g.*, the optimum price and the EOQ, both of which have the effect of maximizing profit.

Ordered Categorisation. Refers to a categorisation in which, although the variable has been not expressible in terms of an underlying measurable variable, it may nevertheless be arranged in order.

Ordered Series. Refers to a set of variate values which have a natural sequence in time or space. In another sense, an ordered series has been a set of variate values which get arrayed in some specific manner related to their values *e.g.*, from the lowest value to the highest value, or from the earliest available value to the latest available value.

Ordinal Number. Refers to a number which denotes order, quality or degree in a group, such as first, second, third.

Ordinal Scale. Refers to a scale with classification and rank. However, equal differences in position on the scale do not necessarily represent equal differences in the variable being used for classification.

Ordinate. Refers to the vertical axis of a graph, the corresponding axis (or base line) being termed as the abscissa, or sometimes as the coordinate axis.

Orthogonal Polynomials. If $P_i \equiv P_i(x)$ has been a polynomial with the coefficient of x^i not zero and $F(x)$ has been a distribution function, then $P_0, P_1, \ldots, P_n$ have been forming a set of

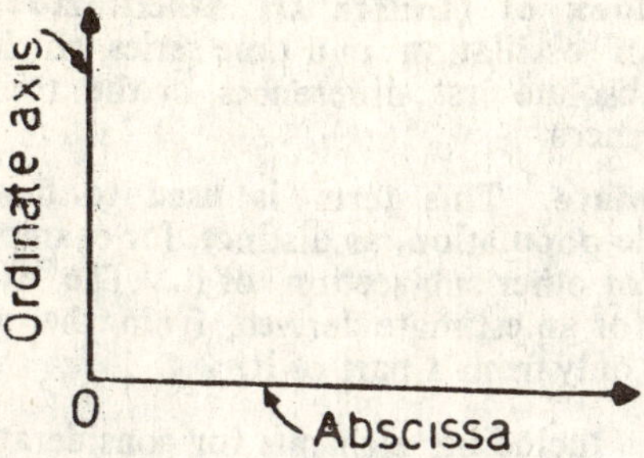

orthogonal polynomials if $\int P_i P_j F(x)dx=0$ $(i \neq j)$ where the integral may include summation of discrete values. If, also, $\int P^2_i F(x)dx=1$ the set has been said to be orthonormal.

Orthogonal Process. Refers to a stochastic process $\{x_t\}$ such that $\xi\{|x_t|^2\}<\infty$ and for which $\xi\{x_s \bar{x}_t\}=0$, $(s \neq t)$ $\bar{x}$ has been the complex conjugate of x.

Orthogonal Regression. For a set of bivariate values (x, y) which are represented as a set of points with Cartesian coordinates (x, y), the so-called 'orthogonal regression' has been straight line such that the sum of squares of perpendiculars from the points on to the line has been a minimum.

Orthogonal Squares. If two latin squares can get super-imposed so that every letter of the first and every letter of the second occupy somewhere the same position the original squares have been said to be orthogonal squares. For example, the two squares :

ABCD		ABCD
BADC	and	CDAB
CDAB		DCBA
DCBA		BADC

combine to yield :

AA	BB	CC	DD
BC	AD	DA	CB
CD	DC	AB	BA
DB	CA	BD	AC

Orthogonal Tests. Refer to another name for test which have been independent, and one which seems better avoided.

Orthogonal System. See orthogonal polynomials.

Oscillation. Refers to an oscillation in a time series or, more generally, in a series ordered in time or space has been a more or less regular fluctuation about the mean value of the series.

Oscillation Index of (Indice Di Oscillazione). Refers to a measure of oscillation in a time series which is obtained as a mean of absolute first differences or the root mean square of first differences.

Over-all Estimate. This term is used to denote an estimate for a whole population, as distinct, for example, from one for a stratum or other sub-section of it. The expression has been also used for an estimate derived from the whole of a sample instead of only from a part of it.

Overcoverage. Including elements for consideration which do not belong to the defined survey population.

Overhead. Of the total annual cost, that part which does not vary with the rate of output, but exists in time. Variously called *fixed costs, oncosts, burden, time costs* and *period costs.* The literature is often somewhat vague about the definition.

Over Identification. See identifiability.

Overlap Design. Refers to an alternative name which was given for serial designs in the context of routine quality control and experiments.

P

Paasche's Index Number. Refers to one of the two basic index numbers of the Laspeyres-Paasche group. It has been similar in form to Laspeyres's index number, but has been currently weighted instead of base weighted. The aggregative formula for a price index is

$$P_{no} = \frac{\Sigma(p_n q_n)}{\Sigma(p_o q_n)} \quad (1)$$

where P_{n_o} denotes the index for the given year, n to the base year 0, = 1.00 ; p denotes the price of a given item, q the quantity of the given item and Σ the sign of summation. As the value, pq, of any item is nearly always much easier to come by than the quantity, q, there would be little point in converting the aggregative form to an arithmetic one, for which the value weight would consist of the quantity in the given period, valued at prices in the base period. Such a conversion would necessitate using the inverse price relative *i.e.*, p_o/p_n, instead of $p_n p_o$. However, an harmonic form is practicable, namely

$$P_{n \cdot o} = \frac{\Sigma(p_n q_n)}{\Sigma\left(p_n q_n \frac{p_o}{p_n}\right)} \quad (2)$$

again necessitating the use of the inverse price relative.

Paasche's quantity index formulae are :

Aggregative

$$Q_{n \cdot o} = \frac{\Sigma(p_n q_n)}{\Sigma(p_n q_o)} \quad (3)$$

Harmonic

$$Q_{n \cdot o} = \frac{\Sigma(p_n q_n)}{\Sigma\left(p_n q_n \frac{q_o}{q_n}\right)} \quad (4)$$

Although Paasche is generally considered to be as good as Laspeyres, it is rarely used for making official index numbers, though owing to the method used for calculating quantity index numbers from price index numbers, *i e.*, by dividing them into the value index as described under Index numbers, the quantity index thus made is Paasche's.

Paasche's index is not necessarily a chain-base index : the four formulae of (1) to (4) above are all fixed-base formulae. Its chain-base form is given under chain-base index numbers.

Applications. Testing Laspeyres's index for bias, and for use as a term in Fisher's ideal index.

Paasche index tends to give a lower value than the Laspeyres index because people tend to shift their purchasing pattern away from commodities whose relative price increase is large and towards those whose relative price increase is small.

Paasche Method. In weighting an aggregates index, the Paasche method uses as weights the quantities consumed during the current period,

Paired Difference Test. A hypothesis test of the difference butween the sample means of two dependent samples.

Paired Sample's. 1. A sample in which the same attribute, or variable, of each member of the sample has been measured twice, under different circumstonces. 2. Two samples in which the members of each have been clearly paired.

Paired-Sample Test. Refers to a test on the differences of the random variable measured in a paired sample, usually of the Null Hypothesis that the mean value of the difference is zero.

Paired Selection Sample Design. Refers to a sample design in which two units have been selected from each stratum. It is a form of replicated sampling.

Paired Selections. The name assigned to two units selected randomly from within a stratum. Paired selections are forming the basis of a paired selection sample design.

Palm's Theorem. It is a limit theorem (1943) which explains why for so many practical situations in which a stationary process has been well approximated by a renewal process, the renewal times have been negative exponentially distributed.

Parabola. An archlike curve described on the surface of a cone by the intersection, at a place below its apex, of a plane parallel to the opposite side. Now generally used of any archlike curve, such as that described by the flight through the air of an arrow or bullet, and that described by a quadratic equation whose coefficient of χ^2 has a negative sign. The graph of a frequency distribution is not called a parabola.

Paradox. A seemingly absurd or self contradictory statement.

Parallel Line Assay. Refers to an important method for the bioassay of a test preparation against a standard. If the expected response gets linearly related to the logarithm of dose the regressions of response on log dose for the two preparations would often be parallel and the distance between them would estimate the logarithm of the relative potency.

Parameter of Location (or Scale). Refers to a parameter of a frequency function which could be identified with some measure of location (or scale).

Parameter Point. If a class of frequency functions is dependent on certain parameters ; *e.g.*, if the unvariate function of x is $f(x, \theta_1, \theta_2, \ldots, \theta_k)$, the domain of variations of the θ's is termed as the parameter space and any particular set of θ's determines a point in that space.

Parametric Hypothesis. A hypothesis which concerns the parameter (*s*) of a distribution.

Parametric Programming. Refers to a development of linear programming in which the parameters of the objective function, formed by the constraints, have been allowed to vary in a determinate fashion.

Parametric Statistics. That brand of statistics which is comprising techniques of statistical analysis and require that the data being analyzed be drawn from a population distribution with a specified form.

Parent Population. Refers to the population from which a sample has been drawn, any member of which could have been selected sometimes this is just referred to a the *population*. See also sample space.

Pareto Curve/Distribution. Refers to an empirical relationship which describes the number of persons y whose income is x. It was first advanced by Pareto (1897) in the form

$$y = Ax^{-(1+a)}, 0 \leqslant x \leqslant \infty.$$

Pareto Index. The coefficient a in the expression for the Pareto curve is generally termed as the Pareto Index. It affords evidence of the concentration of incomes, or, more generally, of the concentration of variate values in distributions of the Pareto type.

Pareto Technique. A method of testing data to discover whether the Pareto principle holds good and of using the data in Pareto form as a basis for control. The Pareto principle states that there is frequently an inverse relation between the percentage of items in each of a set of sub-classes and the importance of the sub-classes.

Pareto-Type Distribution. Refers to a loosely used expression which is used to denote any distribution shaped similarly to that of Pareto. It is probably better avoided unless the distribution has been actually Paretian apart from origin and scale.

Parity. Equality : the term 'parity' is more commonly used in expressing equality of two sets of terms in an equation.

Part Assembly Sequence. When a part method is used in training, there are various ways of assembling the parts so as to ensure that newly learnt parts are integrated with parts learnt previously.

Partial Association. Refers to the measure of association in sub-populations of qualitative characteristics analogous to the partial correlations between quantitative variates. For example, if A, B and C have been three attributes the partial association of A and B with respect to C can be the association of A and B in the sub-population of members bearing the attribute C.

Partial Confounding. If, in a factorial experiment with several replicates, there have been interactions which have been confounded in some replicates but not in others these interactions have been said to be partially confounded.

Partial Correlation. Refers to the correlation between two variates, controlling for one or more other variates.

Partial Regression. In regression analysis, this form refers to the coefficient of an 'independent' variable in the complete regression equation, that is to say, the regression equation which is involving all the variables under consideration. It is termed as partial, not a vary fortunate expression, because it differs from the coefficient which would be obtained if certain, perhaps all, other independent variables get ignored and a simple regression of the dependent variate calculated in the one independent variable.

Partial Serial Correlation. Correlation within a series after permitting the influence of observations within the limits of the lag ; it has been analogous to the ordinary concept of partial correlation, and is also known by the alternative name partial auto-correlation.

Partition of Chi-squared (χ^2). In certain circumstances the sum of squares of standardised Normal variates about their mean, which gets distributed as χ^2, could be divided into two or more parts each of which is also distributed as χ^2 independently of the others. This is termed as a partition of χ^2.

Pascal Distribution. An alternative name for the negative binomial distribution, presumably because some untraced individual thought that Pascal discovered it.

Pascal's Distribution or Negative Binomial Distribution. It is given by :

$$p(x) = {}^{x-1}C_{x-n}\, p^n q^{x-n}$$

for $x>n$ where $0<p<1$ and $q=1-p$. This is the probability distribution for the number of bernoulli trials x needed to achieve n successes. Each trial has probability p of success, q of failure.

The mean and variance of this distribution have been given by :

$$Mean = \frac{nq}{p}$$

$$Variance = \frac{nq}{p^2}$$

When $n=1$, Pascal's distribution becomes the geometric distribution.

Pascal Triangle. A triangle of numbers named after Pascal (1623-62), to whom it gets attributed, showing the coefficients of the binomial expression raised to the power of k($k=1,...n$, where *n* has been the number of rows of the triangle). Assuming the outcomes to be head (H) or tail (T) on tossing an unbiased coin, and *k* to be the number of occasions on which the coin is tossed, the Pascal triangle is shown below.

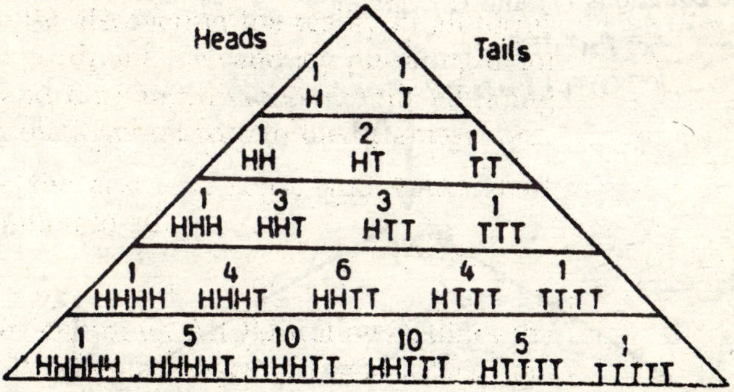

The triangle gives the coefficients where the binomial expression $(h+t)$ expanded *k* times (*i.e.*, $(h+t)^3=h^3+3h^2t+3ht^2+t^3$, etc). Thus if a coin is tossed three times, the likelihood of outcomes is measured by dividing the respective coefficients (or raw elements) by the row total. The likelihood that all outcomes will be head is $^1/_8$, that there will be two heads and one tail is $^3/_8$, that there will be one head and two tail is $^3/_8$ and that all outcomes will be tails is $^1/_8$. Each element of each row is the sum of adjacent elements from the previous row *e.g.*, 1+3=4.

Path Analysis. Refers to a method by which the coefficient of a path model have been estimated. Let us consider the following simple path model.

The equations may be written as :

$$x_2=p_{21}x_1 \qquad (1)$$

$$y=p_1x_1+p_2x_2 \qquad (2)$$

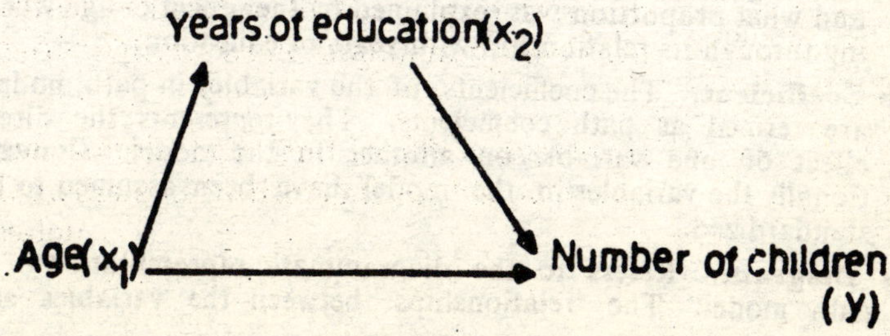

indicating that x_2 (education) has been determined by x_1 (age) and that y (number of children) is determined both by x_1 (age) and x_2 (education). However, the relationship has been not exact—other factors not included in the model will also affect the variables—and residual or error terms should be added to the equations (1) and (2), giving :

$$x_2 = p_{21}x_1 = p_{2u}x_u \quad (1)$$

$$y = x_{y1}x_1 + p_{y2}x_2 = p_{yv}x_v \quad (2)$$

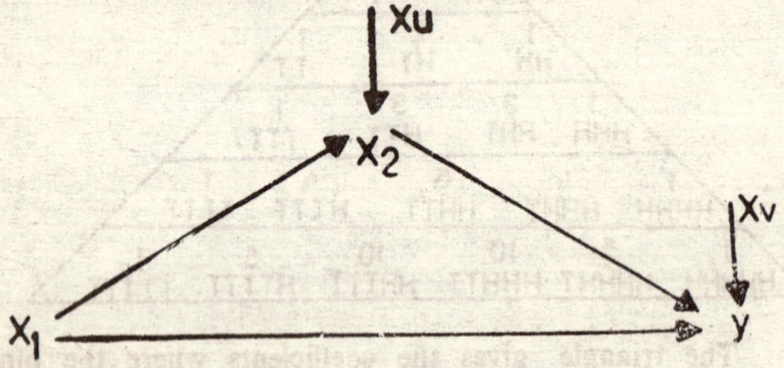

These residuals are shown on the diagram as follows :

The first residual (x_u) represents factors included or ignored in equation (1) ; the second residual (x_v) represents the factors excluded or ignored in equation (2).

The coefficients p_{21}, p_{2u}, p_{y1}, p_{y2} and p_{yv} could be estimated in either of two ways : (*a*) by applying ordinary multiple regression methods to each of the equations (1′) and (2′) ; (*b*) by the method of decomposition of correlation coefficients, first developed by Sewall Wright (1921) and 1934.

Path analysis has been a useful method of measuring the direct influence along each path in the system and thus allocating the variation in a particular variable among the possible causes of that variation. In the example above we would ascertain what proportion of the variation in number of children could be explained directly by the age of the women, and what proportion was explained by the effect of age working through its relationship with years of education.

Path Coefficient. The coefficients of the variables in path modele are termed as path coefficients. They represents the direct effect of one variable on another in the model. Conventionally the variables in the model have been assumed to be standardized.

Path Diagram. Refers to the diagrammatic representation of a path model. The relationships between the variables are shown on the diagrams by lines, two kinds of relationships can be represented—directed arrows indicating a causal relationship in the direction in which the arrow is pointing; and curved lines indicating simply an association between two variables.

Path Model. Refers to a causal model in which the relationships between the variables have been represented by magnitudes of effects along paths connecting the variables in the model. The model could be represented either as a path diagram or as a system of equations.

Pearl-Read Curve. Refers to another name for the logistic curve, a general form of growth curve.

Pearson Coefficient of Correlation. The product moment coefficient of correlation has been sometimes referred to as the Pearson coefficient of correlation due to K. Pearson's part in introducing it into general use.

Pearson Criterion. See criterion.

Pearson Curve. Refers to a distribution from the family of frequency distributions which was developed by K. Pearson. The basic equation of the family is :

$$\frac{df}{dx}=\frac{(x-a)\,f}{b_0+b_1x+b^2x_2}$$

where f denotes the frequency function. The constants of this equation could be expressed in terms of the first four moments, if these exist. The explicit solutions have been classified into types according to the nature of the roots of the equation $b_0+b_1x+b_2x^2=0$.

Pearson Measure of Skewness. Refers to a measure of skewness which was proposed by K. Pearson in the form :

$$\text{Skewness}=\frac{\text{Mean-Mode}}{\text{Standard Deviation}}$$

Peek Group. Refers to a social group which is composed of individuals of approximately the same age or status.

Pentad Criterion. In factor analysis, this term refers to an extension of the tetrad criterion which was developed by Kelley and Holzinger and based on sets of five correlations from the correlation matrix.

Percentage. Refers to the proportion or rate per hundred parts. Symbol : %. To write a number as a percentage, all that is needed is to multiply it by 100.

Percentage Bar Diagram. It is a method of showing a population which has been divided into several sets. In this sort of diagram, a single bar is representing the whole populations (100%) and is divided into parts in the correct proportions (Fig. 1).

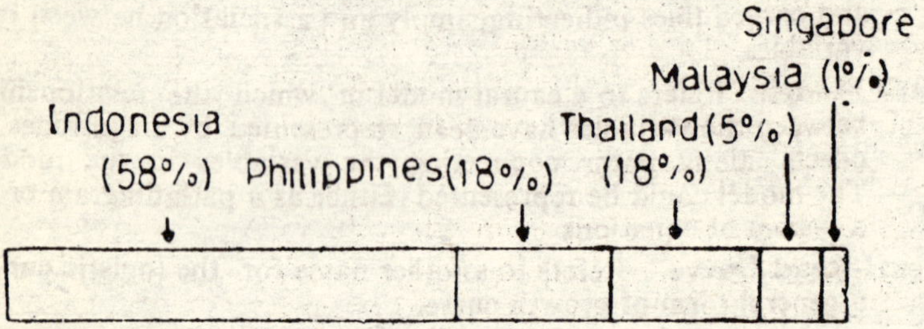

Fig. 1. Percentage bar diagram. Populations of the countries in the Association of Southeast Asian Nations (ASEAN).

Percentage Diagram. Refers to a diagram which shows a simple analysis of statistical data in terms of percentages. The actual form of the diagram could vary ; examples have been the bar chart and the pie chart.

Percentage Distribution. Refers to a frequency distribution having the total frequency equated to one hundred and the individual class frequencies expressed in proportion to that figure.

Percentage Laws. The percentage rate per unit to which a dependent variable tends to rise or fall when an independent variable is doubled. The eighty per cent law said to apply to learning is well known : it states that when the cumulative output of a given product doubles, the labour time per unit of output falls to 80%. The laws apply only to relationships that are logarithmic.

Percentage Point. A level of significance which is expressed in percentage form.

Percentage Relative. Ratio of a current value to a base value with the result multiplied by 100.

Percentage Standard Deviation. A regrettable synonym for the coefficient of variation.

Percentiles. The *k*th percentile is defined as the value of the data such that *k* per cent of the observations are less than that value and 100-*k* per cent are greater than that value. For example, the 50th percentile is called the median.

Perfect Competition. Exists where market forces alone decide price. Even governments cannot exercise control of prices, where competition is perfect, without first making it imperfect by rationing or the control of supplies. In order to have control of their own prices, suppliers have sought to create market imperfection by branding and by forming cartels and trust.

Performance and Cost Indices. A performance index is a ratio of two quantities, one of which represents the output or part of the output and the other the input or part of the input of a system.

A cost index is a special type of performance .index in which the input is a cost and the output may be expressed in monetary units or physical units.

Performance Rating. Performance in behaviour when related to a standard. Performance rating is the rating of the performance of a person or group.

Periodicity. Refers to the characteristic of happening at regular intervals in time. The term not only relates to cyclical fluctuations and seasonal fluctuations, but sometimes to any wave-like oscillation which has identifiable maxima and minima for each of a series of cycles.

Periodic Process. If any realisation of a stationary stochastic process yields a series which has been strictly periodic then the process has been a periodic process.

Period of a State. The period of a return state k of a Markov chain has been the greatest common divisor λ_k of the set of in tegty n for which $f_{kk}(n)>0$, where $f_{kk}(n)$ has been the probabilers that the (first) return time of state k is n.

If λ_k has been greater than unity, state k has been termed as 'periodic' ; if $\lambda_k=1$ the state has been called aperiodic.

Permissible Estimator. Refers to an estimator which yields permissible estimators for all possible samples ; permissible meaning that the estimates lie in the known range of the parameter.

Permutation. Refers to an order in which a given set of objects can be arranged. Thus, for example, the set ABCD may be ordered in the following permutations :

ABCD	BACD	CBDA	DABC
ABDC	BADC	CBAD	DACB
ACBD	BCDA	CABD	DBCA
ACDB	BCAD	CADB	DBAC
ADBC	BDAC	CDBA	DCAB
ADCB	BDCA	CDAB	DCBA

Note that there have been twenty-four arrangements, or permutations, of the set ABCD. It can be expressed as $4!=4\times3\times2\times1=24$.

The number of permutations (of r elements) that can be obtained from a set of objects of size n is :

$$\frac{n!}{r!\,(n-r)\,!}\times r!$$

that is, the number of combinations of size r multiplied by the number of permutations of each r elements. This simplifies to :

$$\frac{n!}{(n-r)\,!}$$

$$\frac{10\,!}{6\,!\times 4\,!}\times 4\,! = 210\times 24$$

$$= 5040$$

or alternatively

$$\frac{10\,!}{6\,!} = 10\times 9\times 8\times 7$$

$$= 5040$$

The term permutation is often misused by those who fill in the football pools and use it to mean combination.

Permutation Tests. Refer to a class of distribution free tests which are based upon the fact that any ordering of a random sample of n items is having the same probability $1/n\,!$.

Persistent State. Refers to a recurrent state for which the average return time has been finite.

Peters' Method. Refers to a method of estimating the standard deviation of a distribution which has been approximately Normal by multiplying the mean deviation by 1.253, this being the ratio $\sqrt{(\frac{1}{2}\pi)}$ appropriate to the Normal distribution.

P-function. This function was introduced by Kingman (1964). If E_t denotes the event that a phenomenon θ occurs at a time $t \geqslant 0$, and

$$\Pr\{E_{t1}\cap E_{t2}\cap \ldots \cap E_{t_n}\}$$
$$= p(_{t1})\,\Pr\{E_{t2-t1}\cap E_{t3-t1}\cap \ldots \cap E_{tn-t1}\}$$

for all positive integers n and $0\leqslant t_1\leqslant t_2\leqslant \ldots \leqslant t_n$, then $p(.)$, with $p(0)=1$, has been termed as a p-function : θ has been described as a regenerative phenomenon that surely happens at $t=0$, and $p(t)=\Pr\{E_t\}$.

Phase. The interval between the turning points of a series which is ordered in time or space is called a phase. The distribution of phase lengths provides one test of random order.

Phase Diagram. A name introduced by Frisch (1937 for a diagram showing two time series x_1 and x_2 plotted as ordinate and abscissa. If the fluctuations of these two variates have been kept in step then the line joining the plotted points will trace a definite pattern, for example similar to an ellipse for oscillatory series.

Phase Spectrum. The sample phase spectrum reveals whether the frequency components in one of two series lead or lag the components at the same frequency in the other series ; it has been a description of the covariance between the two series.

Pictogram. A method of pictorial representation of quantities or magnitudes in a graph, bar-chart, band-chart or cartogram by using a number of miniature drawings of the subject matter, or of related symbols, *e.g.*. £ signs for pounds sterling, barrels for crude oil production, and so on.

Pie Diagram. A more picturesque term for the circular diagram or chart.

Piecewise Regression. Refers to an alternative name for segmented regression.

Pivot. Refers to the essential element in the inversion process which is necessary for the solution of a linear program.

Platykurtic. A slightly peaked distribution.

Plot. In experimental design this term is used to refer to the basic unit of the experimental material. Although it derives from the physical unit of a plot of land in an agricultural trial, its interpretation has been very much more general according to the subject matter of the particular design.

Point Binomial. It is an alternative name for the Bernoulli, or binomial, distribution. The word 'point', which is not necessary, arises from the discrete character of the variate.

Point Biserial Correlation. Refers to a modification of the biserial correlation to the case where one variate, instead of being based on a dichotomy of an underlying continuous variate, has been discontinuous and two valued.

Point Bivariate Distribution. An alternative name for a bivariate distribution of two discrete variates.

Point Density. Refers to the relative frequency, or probability mass, which may get located at the different point values of a discontinuous variate. The term has been better avoided in favour of frequency or probability mass, the word 'density' being reserved for continuous probabilities.

Point Estimate. A single number that is used to estimate an unknown population parameter.

Point Estimation. Refers to estimation of a parameter of a parent population as a single value. This contrasts with *interval estimation*, where the parameter is estimated to lie within an interval.

Point-of-time Statistics. Statistics that relate to a point of time as distinct from rate-of-flow statistics. Point-of-time statistics include stock figures, the book values of fixed assets price, marginal cost marginal revenue and unit cost ; rate-of-flow statistics cover figures that are expressed per unit of time (day, week, year, etc.), such as operating costs, sales, annual costs and fixed costs.

Poisson Binomial Distribution. Refers to a discrete distribution of the number of successes in 'n' independent trials having probability p_j of success in the jth trial. The binomial distribution has been a special case where the parameters are $n, p_1, p_2, p_3, \ldots p_n$ and $p_1=p_2=p_3=\ldots=p_n=p$.

Poisson Distribution. Refers to a special type of probability distribution which can be applied to a large number of physical and natural phenomena. Suppose there have been a large number of occasions during a particular period of time in which some event can occur, and on any one occasion there has been a small but identical probability that it will occur, but the average rate of occurrence for identical periods is the same, then the Poisson distribution would provide the probabilities that during the period of time the event will occur a specified number of times.

Poisson-Normal Distribution. Refers to a distribution which resulted from the expected value of a Poisson distribution, itself taking a truncated Normal distribution. It is also sometimes termed as the Gauss-Poisson distribution.

Poisson-Pascal Distribution. A name which is sometimes assigned to the distribution obtained when a negative binomial distribution and a Poisson distribution have been convoluted.

Poisson Probability Paper. Graph paper which shows curves of the relationship between P and λ where

$$P=e^{-\lambda}\left\{\frac{\lambda^c}{c!}+\frac{\lambda^{c+1}}{(c+1)!}+\ldots\right\}.$$

The axes could be calibrated linearly in P and λ but other scales have been in use.

Poisson Process. Refers to a stochastic process which may be considered as a particular case of a birth process in which the parameter λ_n has been a constant λ.

Poisson Truncated Normal Distribution. Refers to a distribution compounded of half the Normal distribution in $\lambda(\lambda \geqslant 0)$ and a Poisson distribution with parameter λ.

Politz-Simmons Method. Refers to a technique which tries to eliminate the need for call-backs in social surveys by using additional information available for those contracted in the first cell.

Polya Process. Refers to a particular case of a birth process in which the parameter λ_n has been given as follows :

$$\lambda_n(t) = (1+an)/(1+at),$$

a being a constant.

Polychoric Correlation. Refers to an extension of tetrachoric correlation to the case of an m by n table where it may be regarded that the two underlying variates have been jointly normally distributed.

Polyspectra. The spectrum of a time series could be considered as the Fourier transform of its autocorrelations, which are dependent on the product of terms $u(t)u(t+k)$. The transform of the product of more than two terms has been termed as a polyspectrum, *e.g.*, for three terms we are having the bispectrum.

Polytomic Table. Refers to a contingency table having more than two categories in the row and column classifications.

Polytomous. Divided into many parts, or categories ; as distinguished, for example, from dichotomous, which means divided into two parts.

Population. The word population, when used by a statistician, may refer to any specified collection of objects, people, organizations, etc. Some examples of a population are (1) individuals in a country, town or constituency ; (2) animals on a farm ; (3) commercial firms in a city ; (4) record cards in a filing system ; (5) children in a school ; or, indeed, any other predefined set of elements. The population is the parent group.

The term 'population' has displaced the older term 'universe', which itself derived from the 'universe of discourse' of logic. It is practically synonymous with 'aggregate' and does not necessarily refer to a collection of living organisms.

Population, Closed. A population into which there is no immigration and from which there is no emigration. Population change, therefore, is determined entirely by the relationship between the birth rate and the death rate.

Population, Finite. In statistics, an existing (or previously existing) population with a fixed number of cases. A finite population may, but does not necessarily, consist of persons. For example, all the automobile accidents in United States during a given five-year period would be a finite population.

Population, Infinite. In statistics, a theoretical population with an unlimited number of cases. An example of an infinite population would be all the possible tosses of a coin.

Population Profile. A type of bar chart which is used to display the population age structure of a country (or city, region, etc.). The population figures have been divided into age groups, and into male and female. The scale for the bars may be for the actual numbers or for their percentages. The chart in Fig. 1 reveals that, for example, there were 1,80,000 boys aged 0 to 4 living in Nepal in 1976.

There have been three main types of age structure of a given population : each has a distinctive population profile. If children have been defined as under 16 and the aged as 65 and over, these can be categorized as follows (Fig. 1) :

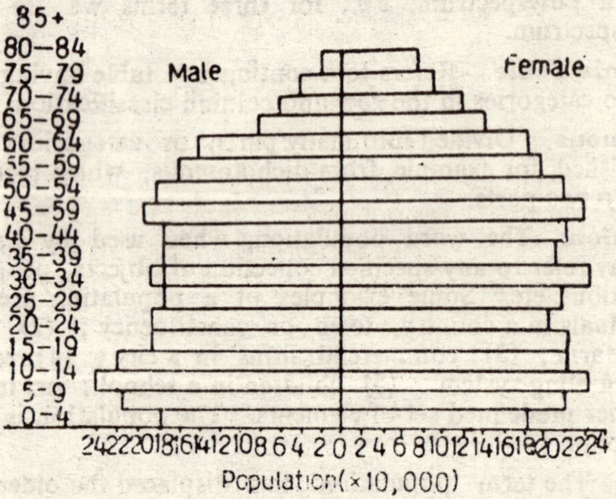

Fig. 1. Population profi'e. The popu'a:ion age structure Nepal in 1976.

Progressive : over 45% children, and under 10% aged.
Regressive : under 30% children, and over 15% aged.
Stationary : 35–40% children, and 10% aged.

An *intermediate* structure has been one changing from one type to another.

Population Standard Deviation. The population standard deviation. σ, is the square root of the population variance. It is a more useful parameter than the variance, because it is expressed in the same units as the data (whereas the units of the variance are the squares of the units of the data).

Positive Binomial Distribution. Refers to the distribution which is formed from the binomial distribution by omitting the value 0. The name is also sometimes assigned to the usual form of the binomial distribution so as to distinguish it from the negative binomial distribution.

Positive Recurrent State. Refers to an alternative name for a persistent state.

Posterior Probability. The term means an estimate of probability after a given event. In contrast, prior probability is the estimate of probability before the event. The terms are relative.

Post-Stratification. A synonym for stratification after selection.

Power. The term has two different meanings :

1. An exponent of logarithm indicating the number of times which a base number must be multiplied by itself to obtain a given value, for example the value 128 is 2^7 (*i.e.*, 2 multiplied by itself 7 times), thus the base number 2 is raised to the power of 7 ; and

2. In statistical testing the power of a test is $1-\beta$, where β is the probability of type II error ; that is, its power is the probability that the alternative hypothesis will be rejected when it is false.

Power Curve. A graph of the values of the power of a test for each value of μ, or other population parameter, for which the alternative hypothesis is true.

Power Efficiency. Alternative name for relative efficiency of a test.

Power Function. The power of a test of the null hypothesis, expressed as a function of a parameter ϕ which expresses the class of alternatives to it. It may be graphically expressed.

Power of the Hypothesis Test. The probability of rejecting the null hypothesis when it is false, *i.e.*, a measure of how well the hypothesis test is working.

Power Statistical. The probability that a given statistical test will reveal a statistically significant relationship when such a relationship really does exist.

Power Sum. Refers to the sum of a series of observations on a variate each of which is raised to the same power. Such quantities are found most frequently in the calculation of moments or similar symmetric functions of the observations.

PPS. A widely used abbreviation for probability proportional to size.

Pre-coded Questions. Questions to which the response has been directly coded into one of a limited number of categories.

Predetermined Variable. In the statistical analysis of models, especially of the economic kind, it is possible to classify the variables as endogenous or exogenous according to whether they represent an integral part of the system or influences impinging on it from without.

Prediction. The term usually implies the estimation of a future value of a variate either (1) by the projection of a trend regression line in time series analysis or (2) by the use of a probabilistic model or (3) both.

Prediction Error of. Refers to the difference between a prediction and the true value of the identity predicted.

Prediction Interval. Refers to the interval between the upper and lower limits attached to a predicted value to show, on a probability basis, its range of error.

Predictive Validity. If a scale has been constructed as a predictor of behaviour or as an indicator of future performance its success (or validity) could be measured by the degree to which it succeeds in this.prediction or indication. This is termed as predictive validity.

Preference Table. The (2^t) paired comparisons of *t* objects, which ties are not permitted, may be displayed by means of 0, 1 variables in a two way table termed as a preference table.

Price Compensation Index. Refers to an index number-of consumers prices which is constructed as a chain index on the basis of a consumer income and varies so as to maintain a constant standard of living. The Laspeyres-Konyus index has been of this type.

Price Discrimination. The charging of different prices to different customers ; usually results from differences in size of order and delivery, and often takes the form of a discount for large orders.

Price Index. Refers to an index number which purports to combine several series of price data into a single series expressing an average level of prices ; *e.g.*, of retail prices or of prices of manufactured products.

Price Index. The ratio of the average of prices in one period or place (referred to as the *given period* or *place*) to the average in another period or place (referred to as the *base period*).

Primary Data. Data which are obtained direct from source and for the solution of a specific problem in hand. Their essential features have been : (1) that they are original, having been obtained by specific investigation rather than from a published, classified or structured source ; and (2) that they are obtained for a specific purpose relevant to the enquiry which has been carried out. The principal means of obtaining primary data have been by self-administered questionnaires, interviews, observation and experiment.

Principle of Equipartition. Refers to the division of the range of a frequency function into a number of parts such that the frequencies corresponding to each part have been equal. This method has been sometimes used in tests of goodness of fit ; it has been also used in sample surveys for the construction of strata.

Principle of the Irrelevant Question. Refers to a technique for getting information about a population which might be damaging or incriminating to individuals.

Prior Probability. (The prior probability of an event A, with respect to event B). Refer to the probability that event A occurs if it is known if event B has occurred or not. This is thus $p(A)$.

The *posterior probability*, for the same pair of events, has been the probability that event A occurs if it has been known that event B has occurred. This is written $p(A \mid B)$.

Priority Queueing. Refers to a queueing system where the actual order of service for the arrivals has been determined by some scheme of relative priority, *i.e.*, not a first come first served discipline.

Probability. This term must be understood from two standpoints, its mathematical foundation and the interpretation of probability measurements.

The mathematical foundation of probability in set theory had been laid by Kolmogorov and others. Probability has been measured in the scale from 0 (=impossible) to 1 (certainty). If two events have been equally likely and one or other of them is certain to happen (such as the outcomes 'head' or 'tail' on tossing an unbiased coin), the probability of each of outcome has been 0.5. Similarly, if a die has been cast, given that it is also unbiased, since it consists of six sides which are equiprobable and it is certain that one side will land uppermost, the probability measurement in respect of each of the outcomes is 1/6. The following scale depicts these four measurements.

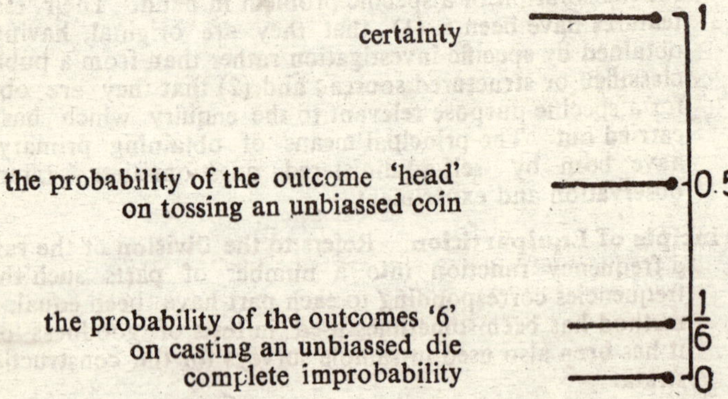

Probability Density Function. Refers to an alternative term for the frequency function when the distribution concerned is conridered as one of probability. It is a mathematical expression showing the relative frequency (or probable density) of x as a function of x, *i.e.*, $f(x)$.

Probability Distribution. Refers to the distribution of the probabilities of occurrence of the values of a variate as distinct from a frequency distribution, that provides the frequencies of the values of the variate.

Probability Element. The probability which is associated with a small interval of a continuous variate and written in some such form as $f(x)dx$, or in general ;

$$f(x_1, x_2, \ldots, x_n)\, dx_1\, dx_2 \ldots dx_n.$$

Probability-generating Function. Refers to an expression which, when expanded, has terms whose coefficients have been the probabilities of the different numbers of occurrences of an event.

The probability-generating function for the binomial distribution with probability p of success, $q=1-p$ of failure, has been given for n trials by :

$$(q+pt)^n$$

where t denotes a success.

The probability-generating function for the Poisson distribution is given by :

$$\sum_{r=0}^{\infty} e^{-\mu} \frac{(\mu)^r}{r!}$$

where μ=mean.

Probability Integral. Refers to an alternative name for the distribution function or the cumulative probability function for continuous variates. For example, the probability integral of a continuous variate x has been a function $F(x)$ having the property that

$$F(a)=Pr\{x \leqslant a\}=\int_{-\infty}^{a} f(x)\, dx$$

where $f(x)$ has been the probability (frequency) function.

Probability Integral Transformation. If x has been a continuous variate having frequency function $f(x)$ and distribution function $F(x)$, a transformation to a new variate y given by

$$y=\int_{-\infty}^{x} f(x)\, dx=F(x)$$

has been termed as the probability integral transformation, y has been uniformly or rectangularly distributed in the range $0 \leqslant y \leqslant 1$.

Probability Limits. Refer to upper and lower limits which are assigned to an estimated value for the purpose of indicating the range within which the true value has been supposed to lie according to some statement of a probabilistic character.

Probability Mass. A term sometimes used for describing the magnitude of a probability or the relative frequency of observations located at a particular variate value, as distinct from being spread over a continuous range.

Probability Paper. Refers to a graph paper which is having the grid along one axis specially ruled so that the distribution function of a specified distribution can be plotted as a straight line against the variate as abscissa. These specially ruled grids have been available for the Normal, binomial, Poisson. log-normal, extreme value and Weibull distribution.

Probability Proportional to Size (PPS). The probability of selection of a unit (usually a cluster of elements) is proportional to its size (the number of elements it contains).

Probability Sampling. Refers to any sampling method in which every element in the population is having a known non-zero probability to selection. In order to be able to use the sample data to make inferences about the population we must use probability sampling. This is because every element must have some chance of selection (*i.e.*, a non-zero probability of selection) or else the characteristics of those elements cannot be represented by the sample, which may give rise to bias in the estimation ; and the probability of selection must be known or else we cannot construct sensible estimators based on the sample responses.

Probability Surface. Refers to a bivariate frequency surface ; that is to say, the three dimensional representation of a bivariate frequency (probability) distribution with the frequency (probability density) alongs one axis and the variates along the other two axes.

Probability Tree. Refers to a graphical representation which is showing the possible outcomes of a series of experiments and their respective probabilities.

Probable Error. A statistical measure of the error that results from the differences between a sample and the total population from which the sample was drawn. The probable error of the mean (P.E.) is the interval in a frequency distribution that centers about the arithmetic mean and includes 50% of all the cases in the distribution, extending from that point which includes half the cases below the mean (25% of all the cases) to that point which includes half the cases above the mean (25% of all the cases). In a normal frequency distribution, this is equal to the quartile deviation.

Probit Analysis. Refers to the analysis of quantal response data which is using the probit transformation.

Procedural Bias. Refers to a bias in a sampling inquiry attributable to the procedure followed in obtaining the information, as distinct from bias due to the use of inferior methods of estimation on the data when obtained.

Process Average. In quality control, this term refers to the proportion of defective items produced by a machine or manufacturing process when it is running normally.

Process Average Proportion Defective. The term is used in quality control to indicate the average proportion of defectives in a batch or sample, and thus the probability that any individual item produced in a controlled process is defective.

Processing Error. Refers to a type of error which can occur in the processing of statistical data. In survey data, for example, processing errors may be including errors of transcription, errors of coding, errors of punching on to cards and errors of arithmetic in tabulation.

Producer's Risk. The risk that a particular batch of satisfactory quality will be rejected by a sampling plan. The risk will depend on the level of quality required and the predetermined criteria of the sampling plan.

Productivity. Unqualified, the output per man hour, man month or man year. Where it refers to a factor of production other than labour, it is usually qualified. The productivity of capital is somewhat ambiguous, and where it refers to the output of a particular type of machine or vehicle, it is probably best expressed in terms of labour productivity.

Product Mix. The body of ratios of the various products made or sold by a company to the total in terms of value. *Optimum product mix* is that body of ratios which maximizes the company's profit. It is a technique of linear programming but, as usually expounded, it makes a number of unrealistic assumptions, *e.g.*, that the company's assets can be switched from producing one type of article to producing any other of the company's range, and that there is no limit to the demand for any product at a fixed price.

Profile Analysis. Refers to a method of clustering the attitude questions in a survey. In common with factor analysis and linkage analysis, it begins from the correlations between the pairs of attitude questions, arranged in the usual correlation matrix. The profile of an attitude question has been its correlation with the other attitude questions, *i.e.*, its row in the correlation matrix.

Profit Planning. Any planning designed to increase or at least maintain the company's profit could be described as profit planning. Howev.r, in the literature is has come to mean an extension of linear break-ever analysis.

Projection. This term finds use in two connected senses. (1) In relation to a time series it implies a future value which is calculated according to predetermined changes in the assumptions of the environment. (2) more recently it is used in probability theory to denote the conditional expectation of a variate.

Propensity. A quantification of a given trend or tendency in terms of proportion, probability or relative frequency. For example, in economic statistics the propensity to save means the proportion of a given income which will be saved, while the propensity to consume means the proportion of a given income which is likely to be consumed. In this case the sum of the two propensities is 1.

Proportional Sub-Class Numbers. See proportional frequency.

Proportionate Stratified Sampling. Stratification represents a procedure by which the population has been divided up into subdivisions, known as strata, in each of which sampling is carried out independently. Proportionate stratified sampling involves selecting a number of units from each stratum which is proportional to the number of units in the stratum ; in other words, we would use the same sampling fraction in each stratum.

Proximity Analysis. A method which is used in numerical taxonomy for arranging items in a line or a plane, or in a space of higher dimensions so that like has been adjacent to like and far from unlike. The method has been based upon rank ordering of inter-point distances and inverse ordering of similarities.

Pseudo-Factor. Refers to an artificial or dummy factor which is used in the design of factorial experiments, generally to render the number of factors a convenient one for the application of some specified balanced design.

Pseudorandom Number. A number taken from a set of numbers which are not truly random, for example, because they are self-repeating or degenerate.

Pseudo-Inverse. Refers to a mathematical device which tries to overcome the fact that a square matrix which is singular does not have an inverse form.

Pseudo-Spectrum. This term is used to denote the mathematical expectation of the estimates of spectral density obtained without regard to the stationarity of the process.

Psi-Square Statistic. Refers to a chi-squared type statistic which can be used as a test of fit with serially correlated observations.

P-Statistics. Refers to a set of statistics which was introduced by S N. Roy (1939) in multivariate analysis. They have been closely allied to the sample values of the characteristic roots of dispersion matrices.

PSU. A widely used abbreviation for primary sampling unit.

Purchasing Power of Money. The quantity of goods that a unit of money will purchase, usually expressed in index number form. To the community in general, it is the inverse of the retail-price index ; to a manufacturer, it may be the inverse of the price index of the materials he purchases and the labour he employs.

Pure Random Process. The simplest example of a stationary process wherein discrete time, all the random variables z_t have been mutually independent. In continuous time the process has been sometimes referred to as 'white noise', relating to the energy characteristics of certain physical phenomena.

Purposive Sample. A sample in which the individual units are selected by some purposive method. It is therefore subject to biasses of personal selection and for this reason is now rarely advocated in its crude-form.

Q

Q-Technique. Refers to a methad of analysis of similarities or relationships in which, given a matrix of *n* observations on *m* individuals, *n* rows and *m* columns, the correlations or other statistical measure have been sought between the *m* columns down the *n* rows *i.e.*, between individuals, as distinct from the R-technique which is looking for relations between variables, namely between rows along the columns.

Quad. Refers to a square-shaped basic cell ; also, the area of such a cell.

Quadrant. One of the four quadrants of the graph. Each quadrant is identified by number or its compass position. The NE quadrant is no. 1 and is the one that is most used as it accommodates the positive values of X and Y :

Sign of X	*Quadrant* Y	
+	+	No. 1 : NE
−	+	No. 2 : NW
−	−	No. 3 : SW
+	−	No. 4 : SE

The horizontal axis is called the *x*-axis, and the vertical axis, the *y*-axis. Their point of intersection is called the origin, and is marked 'O' : it provides the position of the dot for $Y=X=0$.

Quadrant Dependence. If the probability of any quadrant $X \leqslant x$, $Y \leqslant y$ under the distribution $F(x, y)$ gets compared with its independence probability, then the pair (x, y) or its distribution as represented by Lehmann (1966) is positively quadrant dependent if

$$\Pr(X \leqslant x, Y \leqslant y) \geqslant \Pr(X \leqslant x) \Pr(Y \leqslant y).$$

The negative dependence is holding with the inequality reversed.

Quadrat. Refers to an area, usually of vegetation, randomly selected for study. It is normally square in shape (hence the name). The ideal size of a quadrat has been the smallest size that has same number of species as would be contained in a larger one. In a *point quadrat* sampling has been done at the points of a square grid covering the quadrat.

Quadratic Estimator. Refers to an estimator which is based upon some quadratic function of sample values. For example, the standard deviation could be estimated from the square root of the variance, a quadratic estimator, or from the mean deviation or the range, which have been linear estimators.

Quadratic Mean. An infrequently used term which denotes any mean which gets derived by employing the squares of the values for which the mean has been produced.

Quadratic Programming. Refers to a major development of linear programming in which some or all of the constraints and the objective function have been quadratic in the variables.

Quadratic Response. Where the relationship between a quantitative response and the dose metameter has been not linear, but has been of the second degree, the response has been sometimes termed a quadratic response.

Quadrature Spectrum. Refers to the form of spectrum which is able to measure the covariance, between the sine and cosine components, or out of phase components, of a sample time series. If the quadrature spectrum has been denoted by $Q_{12}(f)$ then

$$Q_{12}(f) = -A_{12}(f) \sin F_{12}(f)$$

where $A_{12}(f)$ and $F_{12}(f)$ have been the phase spectrum and cross-amplitude spectrum of the two series.

Qualitative Data. Data that are not in numerical form. Qualitative data are data in the form of descriptive accounts of observations, inverviews, or written material. Data in qualitative form may be possible to quantify.

Quality Control. Refers to the control of the relative quality of a manufactured product, usually by statistical sampling techniques. The reason for using statistical analysis has been that the product has been manufactured in such large numbers that complete inspection of all articles has been impossible Samples from batches have been therefore analysed either ; (1) with the object of discovering the number of defective items ; or (2) with the object of comparing the mean and standard deviation of product size or component quantity with the desirable size or component quantity, given tolerance limits.

Quantal Response. The response of a subject to a stimulus has been said to be quantal when the only observable, or the only recorded, consequence of applying the stimulus has been the presence or absence of a certain reaction *e.g.*, death. A quantal response could be expressed as a two valued variate which is taking values 0 and 1.

Quantiles. Values as certain positions in a set of numbers arranged in order of magnitude. Common examples are deciles and percentiles.

Quantitative Data. Refers to the data which is classified by some numerical value. Compare qualitative data. Thus, data on the number of hours of sleep recorded for certain hospital patients has been expressed numerically, and has been quantitative data.

Quantitative Response. Refers to a reaction, by an experimental unit to a given stimulus, which could be measured on a variare scale. For example, the response could be measured in terms of weight, size or reaction time : in particular the survival time.

Quantity Index. A measure of how much the number or quantity of a variable changes over time.

Quantity Relative. The ratio of the physical quantity produced, sold or in stock of an article in a given period. or at a given time or place, to the quantity in a base period, time or place, *i.e.*,

$$\text{Quantity relative} = \frac{q_n}{q_o}$$

Quantity Weights. When constructing a price index it has been normal to form a weighted average of the relative prices of the items covered by the index. These weights have been the quantities of the items purchased in some reference period, which could be either the current period for which the index has been calculated or some previous base period. Thus the Laspeyres index can be written as

$$\sum_{i=1}^{n} w_i \frac{p_i^{(1)}}{p_i^{(0)}} \times 100$$

where

$$w_i = \frac{p_i^{(0)}\, q_i^{(0)}}{\Sigma p_i^{(0)}\, q_i^{(0)}}$$

i.e., where the weights w_i have been the expenditures of the various commodities purchased in the base period (period 0) expressed as a proportion of the sum of all expenditures ; $p_i^{(1)}$ and $p_i^{(0)}$ $(i=1, 2, \ldots n)$ have been the prices of the commodities in time periods 1 and 0 respectively.

Quantum Changes. Refers to the relative changes in quantities purchased between two periods which are used to construct a quantity index number. They play the same role in the quantity index as price changes play in a price index.

Quantum Hypothesis. Refers to a hypothesis in which the possible values of a parameter have been discrete and hence increase by quantum jumps.

Quantum Index. Refers to an index number which purports to show the changes in quantity, generally of goods or services produced, purchased or sold, independently of changes in prices or money values. One such index has been of the Laspeyres' type obtained by weighting quantities in the given and base period by prices in the base period. Quantum index numbers do not necessarily measure changes in volume or weight.

Quartile. Refers to the value of a variable below which three quarters (1st or upper quartile) or one quarter (3rd or lower quartile) of a distribution lie. The median has been the second quartile.

Quartile Deviation. Refers to a measure of dispersion which is based upon the distance between certain representative values of the variate. In this case the representative values have been the upper and lower quartiles and the quartile deviation is defined by.

Quartile, Lower. In a frequency distribution, the first quartile or 25th percentile, below which lies one quarter of the cases.

Quartile, Upper. In a frequency distribution, the third quartile or 75th percentile, below which lie three-quarters of the cases and above which lies one-quarter of these cases.

Quartile Measure of Skewness. Refers tc an easily obtainable measure of skewness which uses the fact that in a skewed distribution the median does not lie exactly midway between the quartiles. The extant of skewnęss could be assessed by doubling the median and deducting this from the sum of the quartiles, Q_3+Q_1-2M. To transform this measure into relative terms, it becomes necessary to obtain the coefficient of skewness, by dividing by the interquartile range. Thus

$$\text{Quartile coefficient of skewness}=\frac{Q_3+Q_1-2M}{(Q_3-Q_1)}$$

It has been the most common quartile measure of skewness. The measure $Q_3+Q_1-2M/\frac{1}{2}(Q_3-Q_1)$ appears in some texts.

Quartile Variation. This term is alternative to the standard deviation as a measure of variation. If the lower quartile has been Q_1, and the upper quartile has been Q_3, the coefficient of quartile variation, denoted V_q has been given by

$$100(Q_3-Q_1)/(R_3+Q_1).$$

Quasi-Factorial Design. Refers to an experimental design for which a formal correspondence could be set up between the treatments and their combinations and the combinations of the treatments of a factorial set.

Quasi-Markov Chain. Refers to a stochastic process over $N+1$ states identified by the integers 0 to N, for which the Markovian property holds between any two states i and j for which $1\leqslant i, j\leqslant N$. Such a quasi-Markov chain has been said to be of order N. This chain arises in the theory of regenerative phenomena.

Quasi-Maximum Likelihood Estimator. Refers to a set of least squares estimators for multiple equation models where the asymptotic distributional properties of the estimators have been almost independent of the form of the distributions of errors in the equation, so that least squares solutions have been virtually equivalent to maximum likelihood solutions.

Quasimedian. Refers to a generalisation of the concept of a sample median to the arithmetic mean of any pair of symmetrically ranked order statistics ; thus for a sample of size n they may be defined by

$$\theta_i = \begin{cases} (X_{(k+1-i)}+X_{(k+1+i)})2 & \text{if } n=2k+1 \\ (X_{(k-i)}-X_{(k+1-i)})/2 & \text{if } n=2k \end{cases}$$

where $X_{(j)}$ denotes the jth order statistic.

The usual definition of a sample median has been obtained by setting i to zero.

Quasi-Normal Equations. If the normal equations of least squares estimation are having the parameter estimates based upon instrumental variables rather than the ordinary model variables, then the equations would be said to be quasi-normal. 'Normal' in this sense is nothing to do with the Normal' (Gaussian) distribution.

Quasi-Range. This term was given by Mosteller (1946) for the difference $x_s - x_r$, where $1 \leqslant r \leqslant s \leqslant n$, in the ascending order statistics of a sample of n observations. Common practice has been to take the range of $n-2i$ observations, omitting the i largest and i smallest.

Quasi-Random Sampling. The employment of a method of sampling which is not truely random, but where the results will with confidence for practical purposes approximate those which would occur if a truly random sampling process were used.

Questionnaire. A document containing a list of questions, copies of which are sent to a number of people or organizations who are invited to provide the answers, or required to do so under the provisions of a statutory instrument, or under the membership rules of a trade association or similar body.

Questionnaire Design. Refers to a formulated series of questions which are used for getting information on special points.. The function of a questionnaire is measurement, and its specification should state the main variables to be measured.

Queueing Problem. The problem of queues, or congestion, arises in a variety of fields where there has been a service to be offered and accepted rather than a product to be made. In general the problem has been concerned with the state of a system, *e.g.*, the length of the queue or queues at a given time, the average waiting time, queue discipline and the mechanism for offering and taking the particular service. The analysis of queueing problems has been making extensive use of the theory of stochastic processes.

Queueing Theory. This is concerned with the mathematical analysis of problems in which items requiring servicing or items providing service occasionally stand idle, and it is necessary to specify either an arrival rate or a service rate.

or both, in order to achieve the most satisfactory (least costly) solution.

There are three basic statistical figures in most queuing situations : (1) the average number in the system, n ; (2) the average rate of arrival, x ; (3) the average rate of departure, which is equal to the average rate of service, y. There are two broad approaches to the problem : the analystical, which involves the use of mathematics ; and simulation, the Monte Carlo method (see below).

With the analytical approach, the equations below may be required to arrive at a solution. It is assumed that $y>x$, so that there is a tendency for the queue to diminish :

Average number in the system, i.e., in the queue and being serviced :

$$\frac{x/y}{1-(x/y)}=\frac{x}{y-x}$$

Queueing time, in terms of the unit of time in rates of which x and y are expressed :

$$\text{Expected}=\frac{1}{y-x}-\frac{1}{y}$$

$$\text{Average}=\frac{x}{y}-\frac{1}{y-x}$$

Queueing plus service time, average :

$$\frac{1}{y-x}$$

Quick Ratio. A business ratio whose current terms are readily available, usually current assets to current liabilities. Sometimes called the *acid ratio*.

Quintiles. Refer to the set of four variate values which divide the total frequency into five equal parts.

Quota Samples. A sample which is usually of human beings, in which each investigator gets instructed to collect information from an assigned number of individuals (the quota) but the individuals are left to his personal choice.

Quota Sampling. This term describes loosely a set of sampling methods which are used widely by market research organizations in interview surveys. The distinguishing feature of a quota sampling scheme has been that each interviewers workload has been based on a fixed quota of individuals who must be representative of the population in respect of some characteristics. The characteristics usually employed have been age, sex and socio-economic status.

Quotient Regression. Refers to a regression of the form

$$y_t=\frac{a_0+a_1x_{1t}+\dots+a_nx_{nt}}{\beta_0+\beta_1z_{1t}+\dots+\beta_mz_{mt}}+\epsilon_t$$

(Wold, 1966), where x and z have been regressor variables and ϵ is a random residual. The coefficients a and β have been generally estimated by least squares in an iterative procedure.

R

Radix. Refers to the number upon which a system of numbers has been based. For example, the radix of a system of natural logarithms has been (=2·718).

Raising Factor. This term is generally synonymous with one of the meanings of factor, that by which a quantity has to be multiplied to obtain another quantity. It is, however, sometimes used specifically to mean the factor by which a sample value has to be raised to obtain a parameter estimate.

Random Balance Design. Refers to a factorial design which is intended to deal with the case where the number of factors exceeds the number of observations. It has been constructed by selecting treatment combinations at random from the possible set, subject to constraints to ensure balance.

Random Bifurcation. Refers to the manner of growth of a tree or rooted graph when the free end that bifurcates on each occasion has been a random choice from those available.

Random Component. If a magnitude is having a number of parts compounded in some way, *e.g.*, by addition or multiplication, any such part as is a variate has been random component of the magnitude.

Random Error. An error in the deviation between the true value and the observed value or between the true value and its estimate.

Random Impulse Process. Refers to a stochastic process which is describing the linear motion of a particle subject to a series of small impluses which have been random.

Random Numbers, Table of. A printed table listing numbers in random order that may be used in selecting a random sample. The arrangement of the numbers in the table is random in all directions.

Random Sample. Refers to a sample which is selected by a method of random selection. The American term 'probability sample' has been probably preferable as the term random may get me confused with 'haphazard'. The term random sample has been also sometimes used to denote a simple random sample, although this usage is not satisfactory. Random samples have been distinguished from quota samples and any form of model sampling.

Random Series. Refers to a series the nůmbers of which may be considered as drawn at random from a fixed distribution.

Random Standardized Normal Deviates. Refer to a set of random numbers which are converted into standard score (Z) equivalents.

Random Start. In systematic sampling the first sample unit is selected at random from the first set of sampling units—the first sampling interval. The sample is then said to have a random start.

Randomization. Refers to a process of ordering a set of objects or numbers using a process that ensures that every possible order (or arrangement of items) of the set has been equally likely.

Randomisation Test. Refers to any test which is characterised by the test statistic being referred under the null hypothesis, to its distribution on permuting the observations in every equally likely way that could have occurred under that hypothesis.

Randomised Model. Refers to a statistical model, generally an experiment design, where the treatment combinations have been assigned to experimental units by some random arrangement.

Randomised Test. This expression is to be avoided. It is found in the meaning of a test in which the decision to accept or reject depends on the value of a random variable. It should not get confused with randomisation test.

Range. Refers to the difference between the largest and the smallest values of a sample or set of variables.

Range Chart. Refers to a chart which is used in statistical quality control on which the recorded quality criterion is the range of samples. This control chart has been used to maintain a check upon the variability of the quality of the particular product or processes.

Rank Correlation. Refers to correlation of data by rank, rather than by numerical value. This can be measured using Kendall's or Spearman's rank correlation coefficient.

Rank Order Statistics. Type of statistics which gets based only on the rank order of the sample observations, *e.g.*, the rank correlation coefficients. 'Rank order statistics' have been distinguishable from 'order statistics', *e.g.*, the median, the range which have made use of the metric values of the observations.

Rank randomisation Tests. Randomisation tests having the observations replaced by their size ranks before computation of the test statistic. The sample space and null distribution of the test statistic have been thereby standardised, so the null distribution may profitably be tabulated.

Ratio Delay Method. An alternative name for the snap reading method.

Ratio Estimator. Refers to an estimator which takes the form of a ratio between two random variables. Such ratios are usually used in sample surveys. The value obtained by using a ratio estimator has been known as a ratio estimate.

Ratio Scale. (1). Refers to a scale of measurement which has the properties of an interval scale together with an absolute zero or origin.

(2) Refers to a graphical scale in which equal absolute variations correspond to equal proportional variations in the data.

Rational Price Fixing. The fixing of prices that maximize profits. The following formula is used to derive the optimum price, as the rational price if often called :

$$p_m = a \frac{e}{e-1} \qquad (1)$$

where p_m is the optimum price, a the marginal cost, and e the price elasticity of demand. Its derivation is given under marginal revenue.

Raw Data. The data from a survey or experiment in their original form, *i.e.*, before being modified or analysed in any way. Raw data are, however, often assumed to be edited and coded. code.

Raw Score. The score as originally obtained in some psychological, educational or other test.

Recall Errors Memory Errors. In answering a factual question the respondent usually relies to some degree on memory to provide an accurate answer.

Recall Loss. In questions dealing with past events, the respondent may simply have forgotten some events and therefore fail to report them. This has been recall loss.

Recode, Recording. Refers to substituting new codes for the original coding of the data.

Record Check. A method which is used occasionally to check on the accuracy of the information collected in a survey. There may be some external source (record) for some of the information obtained. A check on age reporting may be done by examining the register of births, for example.

Records Tests. Distribution free tests for trend in time which series are based on the breaking of record values.

Rectified Index Number. Refers to an index number formula which has been obtained by taking the geometric mean of two other index numbers of opposite bias.

Rectifying Inspection. Refers to the inspection of a product which aims at removing any defective units found and replacing them by effective units.

Rectilinear Trend. An alternative name for a linear trend.

Recurrence Game. Refers to a sequence of trials of an event which have been conducted as a game, in that a 'reward' has been received, or a 'fee' incurred, under certain recurrences.

Recurrent Markov Chain. A periodic Markov chain.

'Recursive' System. The word 'recursive' is used by some workers to denote relations which have been 'recurrent', presumably under the impression that the former is the correct adjectival form derived from the verb 'to recur'. A purist will avoid the word in this sense.

Reduced Design. Refers to a device in experiment design for restricting the large size of block needed by a full design.

Reduced Equations. Refers to a method of estimation in econometrics whereby the original equations get modified so that each endogenous variable gets expressed as a function of the set of exogenous variables and, possibly, the errors.

Reduced Sample. A term which has been sometimes applied to the involuntary censoring of a sample when some items have been no longer observable.

Reduced Variate. Refers to a standardised variate which is using location and scale statistics other than the mean and the standard deviation.

Reference Period. In one sense this term has been synonymous with base period. It may also refer to the length of time, *e.g.*, week or year for which data has been collected.

Reference Set. An alternative term for fundamental probability set.

Refusal Rate. In a sample survey we attempt to obtain information from all eligible selected sample elements. Some of these may refuse to provide the information (*i.e.*, refuse to respond). The proportion of those eligible elements who refuse is called the refusal rate.

Refusals. Refer to one of the categories of non-response, those who, although successfully contracted, refuse to provide the information requested.

Regressand. A synonym for dependent variable in a regression relation. This term is sometimes used instead of dependent variable in regression analysis.

Regression Coefficient. Is the coefficient or any regressor (or independent variate) in a regression equation. Thus, if the regression equation is simply $Y=a+bX+\epsilon$, the regression coefficient is b.

Regression Curvilinear. A relationship between two (or more) variables in which there is not a constant ratio of change between them.

Rejectable Quality Level. Refers to the level of quality as determined by percentage defective, for example, for which a buyer would wish to have only a low probability of accepting. This probability has been the consumer's risk.

Rejection. In statistics the term rejection is applied to the rejection either of a hypothesis, or in the case of quality control of a set of items of a given quality

Rejection Region. In the theory of testing hypotheses, this term refers to a region of the sample space such that if a sample point falls within it the hypothesis under test gets rejected.

Rejective Sampling. Refers to sampling with unequal probabilities with replacement in which the whole sample has been rejected as soon as any individual has been selected a second time.

Relative Cyclical Residual. A measure of cyclical variation, it uses the percentage deviation from the trend for each value in the series.

Renewal Process. Refers to a class of stochastic processes in which times between events have been independently and identically distributed.

Renewal Theory. Refers to an application of the analysis of recurrent events to problems concerning the duration of life in aggregates of physical equipment. Such aggregates have been sometimes referred to as self-renewing when the failure of any unit results in its replacement.

Repeated Survey. Refers to a sample survey which is carried out more than once with essentially the same questionnaire or schedule but not necessarily with the same sample units.

Replacement Cost. The current cost of materials and bought-out parts and components ; the expected or forecast cost of replacing a wasting fixed asset on life expiry. Both have an important part to play in the calculation of company net profits, especially in periods of inflation and deflation.

Replacement Process. Refers to a sequential control process for which, at various points, the system can be returned to some initial state.

Residual Heading. A heading in a statistical classification designed to render the classification, or a section of it, exhaustive.

Residual Method. A method of describing the cyclical component of a time series.

Residual Sum of Squares. See error sum of squares.

Residual Treatment Effect. In experiments which get continued over several consecutive periods of time on the same individual it has been important to consider whether the effect of the experimental treatments administered during one period has been carried over into the next and subsequent periods.

Residual Variance. Refers to that part of the variance of a set of data which remains after the effect of certain systematic elements such as treatments is removed. It is able to measure the variability due to unexplained causes or experimental error.

Response Metameter. Refers to the transformed measurement of the response to a given stimulus.

Response Rate. In a sample survey an attempt is made to obtain information (the response) from all eligible selected sample elements. Those eligible elements who provide the information are the respondents. The proportion of all those eligible elements in the sample who do respond is called the response rate.

Response Variance. Refers to a measure of the variability of those contributions to response error that tend to cancel out with sufficiently large samples.

Restricted Randomisation. In complex factorial designs the randomisation of the allocation of treatments could not wholly eliminate systematic features which have been felt to be undesirable. The situation may sometimes be met by imposing conditions on the randomisation which is then considered to be restricted.

Restricted Chi-square Test. A modification of the chi square test which was proposed by Neyman (1949) where the restriction is on the alternative hypotheses as derived from an explicit model. The use of this restriction enables one to get higher power in some directions than in others.

Restricted Sequential Procedure. Refers to a class of closed sequential procedure which was proposed by Armitage (1957) incorporating the principle of truncation in order to reduce the variability of sample number.

Retail Price Index. Formerly the cost-of-living index ; compiled monthly and published monthly through the press.

Return Time. In time series, this term refers to the interval of time taken by the series to return to some assigned value, usually an extreme value, as, for instance, 'the return period of flooding' in a river.

Return State. In the theory of Markov chains, this state refers to a state which communicates with itself. A state which communicates with no state, not even itself, is termed as a non-return state.

Revenue Productivity Index. A productivity index that rejects value added as the measure of output, and adopts instead a production index based on the value of production index divided by the price index of the company's output weighted by reference to values in the base period. This would give a currently weighted production index (Paasche).

Reversal Design. Alternative name for switch-back design.

Reversal Test. This term is used in two quite different connections ; (*a*) in certain tests of consistence for index numbers : factor reversal test and time reversal test ; (*b*) in the analysis of time series, where one of the tests for random order has been based upon 'reversals' in the series

Rho. The Greek letter (written ρ), usually used to symbolize the rank difference coefficient of correlation. See correlation, rank difference.

Riemann Distribution. Zipf (1932) posulated the Riemann distribution

$$\phi(R)=[\zeta(s)]^{-1}R^{-s}$$

where $s>1$, $R=1, 2, 3$

as a suitable mathematical representation of word frequency distributions.

Right and Wrong Cases Method. Refers to a method of analysis which was proposed by Muller (1879) for quantal response data arising from psycho-physical experiment.

Right Angular Design. Refers to an association scheme for partially balanced incomplete block designs and defined four-associate class designs.

Risk Function. The word 'risk' finds use in statistics in its ordinary sense and, apart from occurence in acturial statistics, has found one special use in the theory of decision functions.

Robust. A test procedure is robust if it is insensitive to departure from one of the general assumptions on which it is based.

Robustness. Several test procedures involving probability levels dependent for their exactitude on assumptions concerning the generating mechanism, *e.g.*, that the parent variation is Normal (Gaussian).

Root Estimator (of a Mean). When we do sampling from a highly skewed population, the root estimator, $(1-c)\,\bar{x}+c\,\bar{u}^2$ where $u_i=\sqrt{x_i}$, can give substantial gain in efficiency over $\bar{x}$ as an estimator of the population mean (Jenkins *et. al.*, 1973).

Root Mean Square Deviation. Refers to the square root of the second moment of a set of observations which are taken about some arbitrary origin, that is to say, the square root of the mean square deviation or mean square error.

Root Mean Square Error. Refers to an alternative name for root mean square deviation.

Rose Diagram. Refers to a form of diagrammatic representation for grouped angular data.

Rosenbaum's Test. Refers to a nonparametric test for the equality of the scale parameters of two populations known to have equal medians.

Rotatable Design. A k-factor experiment may be defined by Box & Hunter (1957) to be having a dth order rotatable design if the expected yield has been a polynomial of degree d in the k factor levels and the variance of the estimated response is a function of the distance from the centre of the design.

Rotation (Rotazione). There has been said to be functional rotation (*rotazione funzionale*) if the role of dependent and independent variables get interchanged ; *e.g.*, from $y=f(x)$ to $x=f(y)$.

Rounding. Adjusting a number, up or down, to a needed level of accuracy. There have been several ways in which the level of accuracy may be presented, such as to a given number of decimal places, to a given number of significant figures, to the nearest whole number, to the nearest hundred, thousand, etc.

Rounding Error. Refers to the error which get resulted from the process of rounding, *i.e.*, by omitting some end digits and replacing by zeros if required.

'Round Robin' Design. In the case where the same n objects are to be used repeatedly throughout an experiment we are having a special form of resolvable design which may or may not be cyclic.

Route Sampling. A procedure which is similar to line sampling and used in surveys of crop acreage in districts which have been well provided with roads.

Row. In statistics, a horizontal listing of data in a series of categories all of which have a common classification. A statistical table is composed of horizontal rows and vertical columns. Thus in the matrix

$$\begin{pmatrix} 2 & 3 & 4 \\ 7 & 12 & 1 \end{pmatrix}$$

the arrays (2, 3, 4) and (7, 12, 1) are rows and (2, 7), (3, 12) and (4, 1) are columns.

Row Percentage. Refers to the percentage of any particular row in a given cell. For example, if the row total is 400 and a particular cell value is 64, its row percentage is 16%.

Row Total. The total of all elements in a row (*i.e.*, $\sum_{i=1}^{k} a_i$ where $a_1, a_2, \ldots a_k$ have been the elements of the row).

R-Technique. A statistical technique commonly used in factor analysis in which the intercorrelations among a number of tests administered to the same subjects are arranged in a matrix (a tabular arrangement in rows and columns).

Run. In statistics, a series of identical symbols, either numerical or nonnumerical (for example, 6, 6, 6 or $+++$), which may or may not be preceded or followed by different symbols.

Runs Test. A nonparametric statistical test that may be used with a single sample to test whether that sample is random, that is, whether the cases included represent a random sample of the population from which they were drawn.

Rutherford's Contagious Distribution. Refers to a discrete probability distribution which was proposed by Rutherford (1954) where the probability of success at any trial has been found to depend linearly upon the number of previous successes.

S

Sack's Theorem. A theorem given by Sacks (1958). This theorem states under very general conditions that when the Robbins-Munro process has been used with constants $a_n = c/n$ (c a constant), the distribution of $(x_n - \theta)$ is asymptotically Normal with zero mean.

Sample. In statistics, a portion of the total number of cases having a given characteristic (or characteristics). A sample consists of a limited number of cases selected for study from a particular population.

Sample, Accidental. A sample selected purely on the basis of convenience and accessibility without regard to the extent to which it is representative of the population from which it is drawn.

Sample, Area. A type of sample selected on the basis of geographic areas. The total region to be surveyed is divided into geographic areas from which a designated number of areas are randomly chosen. These in turn are subdivided into smaller areas from which a designated number again are randomly selected. Within the finally selected subareas, all or a specified proportion (randomly chosen) of the households are studied.

Sample, Biased. A sample selected in such a manner that certain types of cases have a greater chance of being included than other types of cases. Thus the sample is not representative of the population from which it was drawn, but differs from the population in some systematic fashion.

Sample Census. A slightly unsatisfactory term. The word census has two meanings : (1) a complete enumeration of a population and (2) an enumeration in which only basic or straightforward data are collected.

Sample, Cluster. A sample selected by using clusters or groupings of cases rather than selecting individual cases directly from the total population.

Sample Design. Sample design has two components—the selection process and the estimation process. The selection process consists of the rules, procedures, and operations by which some members of the population are included in the sample. The estimation process consists of the rules and operations for computing sample estimates of the population parameters or population values.

Sample Selection. Refers to the process by which we designate the population elements, which are to be included in the sample.

Sample Size. The number of units included in a sample. It generally refers to the number of elements in the sample. In multistage sampling, for example, the sample size has been the number of final stage units in the sample.

Samples are often called small if their size is under about 30.

Sample Space. Refers to the set of individuals or items available to be sampled ; the parent population from which a sample has been drawn.

Sample Statistic. A term which is synonymous with statistic ; any measure calculated on the basis of the observations in the sample.

Sample, Stratified. A sample selected by dividing a population into categories (strata) on the basis of certain relevant variables and selecting a determined number or cases from each category, thus assuring that each category will be represented in the sample.

Sample Survey. The term used for a survey which is carried out using a sampling method, *i.e.*, in which a portion only, and not the whole population, is surveyed.

Sample, Systematic. A sample obtained by selecting every nth case from an arbitrary list (for example, an alphabetical list) of all the cases in the population.

SAS. A computer program for doing regression and other statistical analyses. Other commonly available packages include MINITAB and SPSS.

Saturated Model. If, in a factorial experiment, θ_i denotes the binomial probability of success for the ith factor combination, then a model with i parameters would be termed a saturated model.

Scalar. A number as distinct from a vector or matrix.

Scale. A standard of measurement consisting of a graduated arrangement of units.

Scale Point. A particular point on a scale.

Scales of Measurement. When a numerical mark or code is allocated to an individual for a particular characteristic, it is necessary to be able to define the meaning of this code in relation to the scores of other individuals in the population.

Scale Parameter. Refers to a parameter of a frequency distribution which gets functionally related to the scale of the variable, *e.g.*, the standard deviation in a Normal distribution.

Scatter. A visual indication of the correlation existing between two series of statistics. The extent to which the dots in a scatter diagram fail to fall into alignment.

Scatter Coefficient. A term given by Frisch (1929) to indicate a property of a multivariate distribution. It has been the square root of the determinant whose elements are the intercorrelations r_{ik} between the pairs of variates ; that is to say, it has been the square root of the correlation determinant.

Scatter Diagram. A diagram showing the joint variation of two variates x and y. Each member is represented by a point whose coordinates on ordinary rectangular axes are the values of the variates which it bears. A set of n observations thus provides n points on the diagram and the scatter or clustering of the points exhibits the relationship between x and y.

Scatter Diagram. A diagram showing a cluster of points representing observations of corresponding values of two variates, such as x and y, such that the two corresponding observed values are co-ordinates of each point on the corresponding axes.

Scedastic Curve. See regression.

Scedasticity. A term, not commonly used, which denotes dispersion. It arises now only in the terms homoscedasticity (equal variances) and heteroscedasticity (unequal variances).

Scenario Writing. A technique that attempts to set up a logical sequence of events in order to show how, starting from the present (or any other given) situation, a future state may evolve step by step. The purpose, however, is not to predict the future (based on Jantsch).

Schach's Two-Sample Tests. Refers to a class of two-sample nonparametric tests for the equality of circular populations, which was proposed by Schach (1969).

Schedule. A set of questions in a sample survey, arranged so as to obtain information from an individual (or set of individuals) on a given subject.

Scheffe's Test. It is a conservative method (1959) which is used for testing the significance of one or more comparisons of mean values arising in analysis of variance where the comparisons get selected by inspection as being of interest.

Schuster Periodogram. Refers to an alternative name for the unqualified term 'periodogram' so-called because it was introduced by Sir Arthur Schuster (1894).

Schwarz's Inequality. Refers to an inequality which in its probabilistic form states that for two arbitrary random variables A and Y

$$\{\xi(XY)\}^2 < \xi(X^2)\xi(Y^2)$$

whenever the mathematical expectations exist. A generalisation of this result for non-negative random variables has been Holder's inequality.

Score. A rating of an individual's performance in quantitative terms or an assessment of some characteristic of an individual on a quantitative scale.

Score, Normative. A score expressed in terms of the population distribution.

Score, Raw. A numerical score in the original form that is obtained from a test.

Scoring, Method Of. An iterative process which is useful for solving non-linear maximum likelihood equations.

Screening Design. Refers to the statistical design for a programme of experiments which has the object of selecting a promising subset of treatments for further development.

Seasonal Variation. Periodic behaviour in a time series whose period is exactly a year is referred to as seasonal variation. It is caused by climatic variation over the year or by some other phenomenon linked to the annual calendar.

Applications. Making seasonal corrections to data such as those used in calculating business ratios, and in budgeting.

Second Limit Theorem. It is a theorem which, states that if the moments of a sequence of distribution functions F_n tend to the moments of a distribution function F, moments of all orders existing, then F_n tends to F, provided that the latter has been uniquely determined by its moments.

Socond Order Stationary. See covariance stationary.

Secondary Data. Data which have been removed at least one stage from source. They are structured or processed in some way.

Secondary Process. In many types of stochastic process it has been possible that the events or occurrences constituting the process may also be characterised by one or more additional random variables. Each of these may be considered as a secondary process.

Secondary Unit. In sampling, a synonym of second stage unit.

Second-Degree Equation. A mathematical form used to describe a parabolic curve that may be used in time series analysis.

Secular Trend. Refers to an alternative name for trend in time series which is sometimes reserved for a trend extending over a long period of years, say, centuries, as against 'trends' extending over decades.

Segmented Regression. Segmented regression has been used when a complete curve to be fitted consists of two or more submodels, and those have to be joined at points whose abscissae are to be estimated.

Selection Bias. There is selection bias in a procedure for sample selection which systematically discriminates against some part of the population and gives either no chance of selection or an unknown chance of selection to that section of the population.

Selection with Equal Probabilities. A sample design in which all the elements in the population or stratum under consideration are allocated the same probability (chance) of being selected into the sample.

Self-Avoiding Random Walks. Refers to a random walk in which the units are having physical size and no two units may occupy the same region of space ; or, on a lattice, a walk which at no point crosses or touches itself.

Self-Correlation Coefficient. Refers to an alternative term for a reliability coefficient, but one which is better avoided.

Self-Conjugate Latin Square. Refers to a Latin square which remains the same if its rows and columns have been interchanged, *i.e.*, it is symmetrical about its main diagonal.

Self-Weighting Estimator. Refers to an estimator which, for the sample design' being used, gives equal weights to the sample observations.

Semi-Average, Method of. It is a particular case of the method of selected points in which the data have been divided into two equal groups and a straight line drawn through the mean of the groups or two other representative points, one in each group. This method has been used to provide a rapid estimate of a linear regression line.

Semi-averages, Method Of. A method of estimating trend. The data are in the form of a time series ; the figures are divided chronologically into two groups, and the means of each taken. These means are then used to estimate the trend.

Semi-Interquartile Range. Refers to an alternative name for quartile deviation that is to say, one half of the distance between the two quartiles of a sample or a distribution.

Semi-Invariant. In older usage this term (introduced by Thiele 1889)) related to what are now called cumulants. The words 'semi-invariant' or 'seminvariant' have been now better confined to statistics which are independent of the origin and are multiplied by a scale factor under transformations of scale.

Semi-Latin Square. Refers to an experimental design for $2k$ treatments arranged in the form of a rectangle with k rows of $2k$; each row being an arrangement of the $2k$ treatments, each pair columns 1, 2, 3, 4, etc., being also an arrangement of the $2k$ treatments ; no row or column therefore containing the same treatment more than once. It may also be considered as a $k \times k$ Latin square with each plot split. The design has been criticised on the grounds that it leads to biassed estimates of error.

Sequential Test. Refers to a test of significance for a statistical hypothesis which is performed by using the methods of sequential analysis. An example has been the sequential probability ratio test.

Sequential Testing. A technique of quality control used where testing is expensive, *e.g.*, where it involves destruction in testing for length of life.

Serial Cluster. Refers to a type of cluster which is used in India in which the actual demarcation of a cluster of listing of units constituting a cluster gets avoided by means of a rule which makes use of the serial numbers already assigned to the units in the frame.

Serial Correlation. The correlation between members of a time series and members of a corresponding time series lagged by a fixed interval of time.

Series Queue. Refers to a queuing system wherein each arrival unit is served in turn by facilities 1, 2, ... , k which are parts of one system. A somewhat different concept has been expressed by the term 'tandem queues'.

Set. Refers to a collection of numbers, objects, etc., that usually have at least one common property or characteristic. The members of a set are termed as *elements*.

Set-Theoretic Model. The model of the set used as the basis for the mathematical analysis of data. In essence, use of this model means that the data are divided into sets or categories and subcategories on the basis of some classification (types of individuals, types of responses, types of experimental situations, group membership, etc.).

Shape Parameter. A term used earlier to denote a parameter of a frequency distribution which is associated with skewness or kurtosis.

Sheppard's Corrections. Refers to the calculation of moments from a grouped frequency distribution introduces certain errors because of assuming that frequencies have been concentrated at the central values of the class intervals.

Shock and Error Model. Refers to a system of equations which is having both stochastic elements associated with specific variables (errors in variables) and elements associated with specific equations in the system, *i.e.* shocks (errors in equations).

Shock Model. In econometric analysis, this term refers to a system of equations which is having random disturbances, as opposed to one in which the variables are subject to errors of observation.

'Short' Distribution. Refers to the convolution of a Neyman type A distribution with parameters λ, θ, and a Poisson distribution with parameter ϕ.

Short Term Fluctuation. Refers to a fluctuation in a time series which is having a short duration ; a continuing set of such fluctuations. 'Short' for this purpose has been a somewhat arbitrarily defined expression.

Sigmoid Curve. Refers to a curve which is lying between two horizontal asymptotes representing a function which increases monotonicity and has a point of inflexion somewhere near a point half-way between them ; hence a curve somewhat resembling a letter S. In statistical work this type of curve is met in connection with, *inter alia*, the distribution function of unimodal distributions, growth curves, such as the logistic, and a particular dose-response relationship in biological assay.

Significance. The term is used mainly in the context of hypothesis testing. It is said to exist when a value lies outside an acceptable interval (known as a confidence interval) and when, consequently, there are good reasons for rejecting a null hypothesis.

Significant Figures. 1. The figures of a number that express its magnitude to a specified degree of accuracy. 2. The number of such figures.

Significance Level. In hypothesis testing it is usual to obtain from a given set of sample data a test statistic calculated for the purpose of the test. This test statistic can only be used if its distribution under the null hypothesis is known.

Significance, Statistical. The probability that an observed sample characteristic or relationship, or difference between samples, represents a true population characteristic, relationship, or difference and is not merely due to chance sampling variation.

Sign Test. A nonparametric statistical test that may be used when data are in the form of paired observations divided into a dichotomized ranking, that is, divided into plus and minus categories indicating more and less or higher and lower.

Similar Action. The name assigned to the action of mixtures of stimuli, *e.g.*, the toxic effect of a mixture of poisons, when the stimuli have been statistically independent and additive. The effect of a mixture is then predictable from the relative proportions of the constituents and the known response of each.

Similarity, Coefficient Of Any measure used to describe the closeness or similarity of elements, groups or variables. The term is used particularly in cluster analysis. The most commonly used to coefficient of similarity is the product-moment correlation coefficient.

Similar Regions. In the theory of testing hypotheses a region in the sample space has been said to be similar to another if a correspondence can be set up between them such that the probability of a sample point falling in a part of one is proportional to the probability that a sample point falls into the corresponding part of the other.

Similarity Index. If two individuals each bear the value of $p(0, 1)$ variables, and in m cases they both exhibit, or do not exhibit the same variate value, the ratio m/p is a similarity index. The complementary quantity $1-(m/p)$ has been a dissimilarity index, and has been used particularly in Cluster Analysis as a distance function.

Similarity Matrix (Matrix of Similarities). Refers to a matrix which presents the coefficients of similarity for each pair of elements in the data set.

Simple Abnormal Curve. See abnormal curve.

Simple Hypothesis. Refers to a statistical hypothesis which completely specifies the distribution function of the variates concerned in the hypothesis.

Simple Lattice Design. See square lattice.

Simple Sample. A random sample is said to be simple when the probabilities of selection of members have been all equal and have been constant throughout the drawing.

Simple Table. A table which is showing only classifications according to one variate or at the most two variates.

Simplex Designs. A simplex has been the *n*-dimensional analogue of the triangle in two, or the tetrahedron in three, dimensions.

The superposition of two or more simplexes to give certain kinds of rotatable designs is known as a simplex sum design.

Simplex Method. Refers to a method of solving a mathematical program by the process of iterative application of rules associated with matrix inversion until an optimal solution gets reached.

Simplex Technique. The general linear programming technique (see also transportation technique).

Simulation. 'The representation of one system by means of another. In particular the representation of physical phenomena by computers, other equipment or models, to facilitate the study of such systems or phenomena, to train operators, etc.'

Simulation Model. Certain types of dynamic model problems cannot be solved by ordinary analytical methods because conditions change rapidly or because of the large number of variables. In such circumstances, it is possible that the problem can be approached by assuming the likely effects of the conditions and producing a computer model which will show the effects of changing one or more of the variables. For example, computer models of business firms can be constructed to simulate the likely effects of trade recessions, seasonal fluctuations, theft, weather conditions, alteration of interest rates, etc.

Simulator. Refers to a physical system which is analogous to a model under study as, for example, an electric network in which the elements have been in correspondence with those of an economic model. The variables of interest in the model appear as physical variables such as voltages and currents and could be studied by an examination of the physical variables in the simulator.

Single Linkage. Refers to a form of cluster analysis. This principle on which it operates is as follows. First, the two elements with the highest similarity in the set have been linked together. The next largest value in the matrix identifies the next pair to be linked. The principle used in the process has been that an element is linked to a cluster if it is sufficiently similar to any of the elements in the cluster. This is why the method is sometimes referred to as nearest neighbour analysis—an element is included in the cluster to which its nearest neighbour belongs. The process continues until the required number of clusters is formed.

Single Sampling. In evaluating the quality of a product or material by inspecting a part rather the whole, single sampling refers to the method in which the decision whether to accept or reject the batch is taken after the inspection of a single sample. This may be contrasted with sequential sampling, where successive samples (elements or groups of elements) are drawn until sufficient evidence is available to make a decision.

Single Sampling Scheme. In quality control, a sampling scheme where only one sample per batch gets inspected. If the number of defectives found is not more than the acceptance number, the batch is accepted. Otherwise it is rejected, or subjected to 100% inspection.

Single Sampling Plan. In quality control, this term refers to a procedure which provides for the drawing of one sample, on the basis of which a lot is accepted or rejected, as distinct from a double sampling plan, which may allow for indecision on the first sample to get resolved by a further sample.

Single Tail Test. An alternative term for a one sided test.

Size of a Region. In the theory of testing statistical hypotheses, the size of a critical region has been a measure of probability and has been the same as probability of an α-arror or error of the first kind. For composite hypotheses it is sometimes used to denote the limits of the α-error where no similar regions exist.

Size (of a Test). The size of the test of a statistical hypothesis (H_0) refers the probability (α) that the hypothesis gets rejected when it is true.

Skew Regression. An obsolete term for curvilinear regression.

Skewness. A form of asymmetry in a unimodal frequency distribution. The distribution is positively skewed when more than 50% of elements lie below the mean and negatively skewed when more than 50% of elements lie above the mean (*i.e.*, less than 50% lie below it).

Slippage Test. Refers to a significance test of k samples in which the hypothesis has been one of homogeneity in the means, as against the alternative that one member or set of members has 'slipped' away from the others.

Slit Chart. See pie chart.

Slope. A constant for any given straight line, whose value represents how much each unit change of the independent variable changes the dependent variable.

Slope Ratio Assay. Refers to a general class of biological assay where the dose response lines for the standard test stimuli have been not in the form of two parallel regression lines but of two lines with different slopes intersecting the ordinate corresponding to zero dose of the stimuli.

Slutzky Process. A synonym for the moving average process.

Slutzky-Yule Effect. Refers to an effect in the averaging of random series which was studied independently by Slutzky and Yule.

Small Numbers, Law of. A term introduced by von Bortkiewicz (1898) to describe the behaviour of rare events obeying a Poisson distribution.

Small Sample. A sample from which, because of its small size, conclusions have to be treated with particular caution. Hypothesis testing may involve the use of special tests.

Smirnov-Birnbaum-Tingey Distribution. Refers to the probability distribution $P_n(\epsilon)$ that if an empirical distribution function $F_n(x)$ has been constructed from a random sample of size n from a continuous probability density having distribution function $F(x)$, then $F(x)$ will be nowhere greater than $F_n(x)+\epsilon$. The earliest attempts at deriving the distribution appeared in 1939 (Smirnov ; Wald and Wolfowitz) ; the explicit form

$$P_n(\epsilon)=1-\epsilon \sum_{j=0}^{[n(1-\epsilon)]} \binom{n}{j}\left(1-\epsilon-\frac{j}{n}\right)^{n-j}\left(\epsilon+\frac{j}{n}\right)^{j-1},$$

where $0<\epsilon\leqslant 1$ $[x]$ implies the integral part of x, is due to Birnbaum and Tingey (1951).

Smooth Test. Refers to a test of goodness of fit between data and hypothesis in which the alternate hypotheses have been considered as moving away from the null hypothesis 'smoothly' in the parameters, *i.e.*, with high continuity and differentiability in them.

Smoothing. Refers to the process of removing fluctuations in an ordered series so that the result shall be 'smooth' in the sense that the first differences have been regular and higher order differences small.

Snedecor's Check. An approximate method of checking standard deviation calculations in normal cases. Briefly, it involves dividing the range by 2 if n (the number of values) is less than 7 ; by 3 if it lies between 7 and 20 ; by 4 if it lies between 21 and 40 ; by 4.5 if it lies between 41 and 80 ; by 5 if n approximates 100 ; and so on.

Snedecor's F-Distribution. Refers to a popular term for the variance-ratio distribution, so-called by Snedecor after the initial of R.A. Fisher's surname.

Snowball Sample. A sample in which additional elements (units) are generated by the units first obtained in the sample.

Snowball Sampling. Refers to a form of sampling which was proposed by Goodman (1961) whereby a random sample of n individuals is drawn from a finite population. Each individual has been asked to name k further individuals : and so on for s-stages. The initial sample solution and its size govern the kind of analysis and inferences that could be drawn from the data.

Sojourn Time. The length of uninterrupted time that a stochastic system spends in a given state.

Spatial Point Process. Refers to a stochastic point process in two dimensions usually represented on a plane.

Spatial Systematic Sample. Refers to a systematic sample taken in two dimensions ; also termed 'plane sampling' by Quenouille (1949).

Spearman Estimator. Refers to the estimator, non-parametric, of the mean effective dose which is relevant to the Spearman-Karbar Method of bio-assay or other stimulus experiment.

Spearman's Rho (ρ). A rank correlation coefficient defined as the product moment correlation coefficient between two sets of rankings of a collection of objects or individuals.

Spectral Average. Refers to the average over a spectral density function $f(w)$ of the form

$$J(A) = \int A(w) f(w) dw$$

where $A(w)$ has been a suitably chosen function. It is found particularly in the smoothing of the spectral density function and $A(w)$ is sometimes termed as the spectral window. Special forms of such a window are known by the names of Daniel, Bartlett, Blackman-Tukey and Parzen.

Spectral Density. Refers to the derivate of the spectral function.

Spectral Distribution Function. See spectral function.

Spectral Function. A necessary and sufficient condition for $\rho(\tau)$, $\tau=0, 1, 2......$, to be an autocorrelation function of a discrete stationary stochastic process has been that it has been expressible in the form :

$$\rho(\tau)=\frac{1}{\pi}\int_{0}^{\pi} \cos\tau w dF(w)$$

where $F(w)$ denotes a non-decreasing function called the spectral distribution function with $F(0)=0$, $F(\pi)=\pi$.

Spectral Weight Function. A weigh function which is used in the estimation of the spectral density : individual specifications are proposed, for example, Daniel (1946), Bartlett (1948), Blackman & Tukey (1959) and Parzen (1961).

Spencer's 15-Point Moving Average. Refers to an example of a moving average having unequal weights. It has been used to smooth or extract the trend from a time series. The smoothed series at time t is a weighted average of past and future observations with weights that are symmetric. Algebraically a symmetric moving average of length 15 can be written

$$S_t=\sum_{j=-7}^{+7} a_j x_{t-1}$$

where (for symmetry) $a_j=a_{-j}$, $j=1, 2, ..., 7$, x_{t-j} is the observation in the series at time $t-j$ and S_t is the moving average or smoothed series.

Spurious Correlation. The introduction into a correlation or regression analysis of a theoretically irrelevant factor, or the introduction of two or more factors as independent variables which, in effect, provide measures of the same force.

Square Root Transformation. Refers to a variate transformation which has been used to 'stabilise the variance' of sample data drawn from Poisson populations ; that is to say, to give variates whose variance would be nearly independent of their means.

Squariance. A term which was proposed by Pitman (1938) in place of the phrase 'sum of squares about the mean' for the sake of brevity.

SRS. A commonly used abbreviation for simple random sampling.

Stability Test. Refers to a rough but convenient test for binomial variation in data which accrue over a period. The cumulative proportion of successes has been plotted against the number of observations. If the variation has been binomial (Bernoullian) the graph of successive points 'settles down' to a straight line, or nearly so, with diminishing fluctuation about such a line.

Stabilisation of Variance. Refers to the process of transforming a variate whose distribution has been dependent on a parameter, so as to make the variance of the transformed variate as insensitive as possible to the values of the parameter.

Stable Law. A distribution function F is termed as stable if to any positive a_1 and a_2 and any real b_1 and b_2, there correspond constants $a>0$ and b such that the relation

$$F(a_1x+b)^* F(a_2x+b_2)=F(ax+b)$$

has been satisfied, where* denotes convolution.

Stable Paretian Distributions. Refers to a class of distributions discussed by Levy (1925) and Mandelbrot (1960) which follow the 'weak' or asymptotic form of the Pareto law having distribution function of the form

$$1-F_x(x)\sim\left(\frac{x}{k}\right)^{-a} \text{ as } x\rightarrow\infty$$

where

$$0<a<2.$$

The distributions have been also termed as stable non-Gaussian distributions.

Stable Process (Distribution). Refers to an alternative term for stationary (stochastic) process. A distribution law has been said to be stable if the convolution of two variates which follow it has been also distributed in the same form : *e.g.*, the sum of two independent Normal variates has been Normal and hence the Normal law has been stable in this sense.

Stable State. Refers to a state of a Markov chain in continuous time for which the expected sojourn time has been greater than zero.

Stacy's Distribution. Refers to a form of generalised T, distribution.

Staircase Distribution. Refers to a picturesque but unnecessary name for a discontinuous distribution which is based upon the fact that its distribution function is a step function.

Staircase Method. Refers to an alternative name for the 'up and down' method especially when used in connection with fatigue tests.

Standard Score. Refers to an alternative name for *z*-score.

Standardised Deviate. Refers to the value of a deviate which is reduced to standardised form (zero mean, unit variance) by subtracting the parent mean and then dividing by the parent standard deviation.

Standardised Mortality Ratio. Refers to an index number in the Paasche form used in the analysis of vital statistics. The ratios of the age specific death rates for the given year to similar rates for the base or standard year have been weighted by the 'expected' deaths in the given year.

Standardised Variate. Refers to a variate in standard measure.

State of Nature. Future event not under the control of the decision maker.

Stationary Distribution. A phrase which has been sometimes used to denote a distribution, *e.g.,* of a human population which remains constant in time. There has been some danger of confusion with a stationary stochastic process and perhaps the expression has been better avoided.

Stationary Population. See stationary distribution.

Stationary Process. A stochastic process is said to be stationary if its statistical properties are not changing over time.

Statistic. A mathematical value that summarizes a characteristic of a sample. A statistic may be a measure of central tendency, of dispersion (see dispersion, statistical), of correlation, of a difference between two samples, or any other summarizing measure based upon sample data. It is used to estimate a parameter of the population from which the sample was drawn, and is sometimes referred to as a parameter estimate. Since most social research deals with samples rather than entire populations, most research yields a statistic rather than a parameter.

Statistical Dependence. The condition when the probability of some event is dependent upon, or affected by, the occurrence of some other event.

Statistical Independence. The condition when the occurrence of one event has no effect upon the probability of occurrence of any other event.

Statistical Inference. The process of inferring generalizations about a population on the basis of sample data. Statistical inference includes the estimation of population parameters and the testing of statistical hypotheses about populations, using data obtained from samples. Statistical inference is based upon probability theory.

Statistical Model. A set of assumptions underlying a statistical test. Every statistical test assumes that certain conditions are met. Usually these assumptions refer to the form of the data, the nature of the sample, the nature of the population from which the sample was drawn, and the character of the variables being studied. The test is valid only if these conditions are met, and thus it is important to know the statistical model on which a test is based before that test is used.

Statistical Progression. A time series with a seasonal pattern or a secular trend or both. A *progressive average* is the AM based on a cumulative total ; it is not a moving average.

Statistical Schedule. A device which sets out the relevant figures of a dependent variable for ranges of the independent variables often based on a regression equation.

Statistical Table. A systematic arrangement of numerical data into a series of rows and columns in order to facilitate the comprehension and comparison of the data. The presentation of data in a statistical table involves the classification of the data into the categories or class intervals specified in the column and row headings.

Statistical Test. A procedure used with quantitative sample data to estimate the probable truth of a hypothesis about the population from which the sample was drawn. The formulas used are based on certain principles, assumptions, and rules of mathematical logic. A wide range of data.

Statistical Tolerance Interval. Refers to the interval between an upper and a lower statistical tolerance limit.

Step Down Procedure. Refers to a procedure (see Ray, 1958) whereby differences have been examined in one observed variable first and then in another observed variable after eliminating by regression the effect of the first, and so on. The variables have been in some sense, arranged in a descending order of importance.

Stemplot. See stem-and-leaf plot.

Stephan's Iterative Process. Refers to an iterative process which was proposed by Stephan (1942) for adjusting sample frequency tables when the expected marginal totals are known.

Stepwise Multiple Regression. A method of eliminating or including variables in regression analysis by means of a stepwise procedure.

Stepwise Regression. Refers to a method of selecting the 'best' set of regressor variables for a regression equation. It is carried out by introducing the variables one at a time (stepwise forward) or by beginning with the whole set and rejecting them one at a time (stepwise backwards).

Stochastic Convergence. Refers to one of several types of convergence concepts in Probability. A sequence $\{x_n\}$ of variates is regarded to converge stochastically to a variate x if

$$\lim_{n \to \infty} \Pr\{|x_n - x| > \epsilon\} = 0$$

for every $\epsilon > 0$. This is also termed as convergence in probability and convergence in measure.

If x_n and y_n have been variates and, as n tends to infinity

$$\Pr\{|x_n - y_n| > \epsilon\}$$

tends to zero x_n is considered to converge stochastically to y_n; but by convention the case where x_n and y_n do not converge has been usually excluded.

Stochastic Dependence. Refers to the relationship between variates which are not independent (independence); as contrasted with mathematical dependence, which has been a relationship between variables. In statistical usage either form has been often referred to as 'dependence', the meaning generally being clear from the context.

Survey. A study or investigation of a population, usually human beings or economic, social or political institutions. The term may be taken to refer to complete coverage of the population as in a census, but it is often used to refer to a study dealing only with a sample from the population (*i.e.*, a sample survey) where the sample observations are used to make inferences or draw conclusions about the whole population.

Survey Method. All methods, procedures and techniques which are used in carrying out social surveys are subsumed in the term survey methods. Thus the term is covering questionaire design, sample design, methods of data collection and methods of data analysis.

Survey Population. The population actually covered by a sample survey. Idealy the survey population should be identical to the target population for which results are desired, but, mainly due to non-coverage and non-response, there may be differences between the two. Other modifications may be necessary simply to make the survey practicable. For example, individuals living in institutions are frequently excluded from consideration on practical grounds.

Survey Sampling. It deals with methods of selecting a sample from a population so as to use the sample observations, together with information on the method of sampling, to make references about the whole population.

Survivor Function. In a renewal process, if x represents the identically distributed times between events in the process, *e.g.*, life times and $F(x)$ has been the distribution function associated with x, the survivor function $R(x)$ has been defined as the complement of $F(x)$. It is a quite general concept associated with the analysis of the life of human, biological or material units.

Syllogism. A common kind of deductive reasoning which, in its complete form, consists of two permises and a conclusion, and comprises three terms, *e.g.*,

All men are mortal.

All kings are men.

Theretore, all kings are mortal.

One term *men*. appears in both of the permises, and is called the middle term. The term *mortal* is called the major term, and it appears in the first or major premise and the conclusion. The term *kings* is called the minor term : it appears in the minor prèmise, and forms the subject of the conclusion. It will be seen that if the two premises are true, the conclusion is also true.

System. The term system is generally used for something which has in it an element of design, which can be represented by a model. Thus a model is a representation of reality, whereas a system comprises that set of states or processes in reality which the model is designed to represent.

Systematic Sample. A sample obtained by the method of systematic sampling. The term is usually confined to samples drawn by selecting units at equally spaced intervals—the sampling interval. The first sample unit is selected from the first sampling interval by some random method.

Systematic Sampling. Refers to a sampling in which each member of the sample is chosen by ordering the frame in some way, and then selecting individuals at regular intervals. Thus a 1% sample of telephone subscribers could be taken by selecting the 1st, 101st, 201st (and so on) of the names listed in the telephone directory.

Systematic Sequare. Early attempts at creating experimental designs for the elimination of variability in two directions orthogonal to each other did not generally employ the principle of randomisation. Hence the allocation of treatment to the rows and columns proceeded in some 'systematic' way. For example, a square design could be laid out as follows with systematic arrangement in the N.W.-S.E. diagonals :

A	B	C	D
D	A	B	C
C	D	A	B
B	C	D	A

Systematic Statistic. A term proposed by Mosteller (1946) for a statistic which consists of a linear combination of order statistics. There appears to be nothing in order statistics or in linearity of combination to warrant the restriction of the word 'systematic' to such quantities, although it has been useful to have a word for them.

Systematic Variation. A term which has been used in two slightly different sense : (*i*) to denote variation which has been deterministic, as opposed to stoachastic, and which can therefore be represented by a deterministic mathematical expression ; and (*ii*) to denote variation in observations resulting from experimental or other situations as a result of factors which have been not under statistical control.

T

Table (Tabulation). It is a systematic summary presentation of data. Tables may be :

1. Single-column univariate frequency tables, and single column category tables showing the frequencies or relative frequencies of a number of dissimilar categories.

2. Bivariate tables.

3. Statistical tables giving critical values of a statistic, such as t, F, χ, ρ, etc. This usage has been very common.

Table Twisting. A tabulating process designed to present figures in the most effective way, either as information, or in furtherance of an argument. The process consists of twisting the table so that the stub (*i.e.*, the line headings) becomes the column headings, and vice versa.

Tabular Presentation. The systematic presentation of statistics.

Tail Area (Of a Distribution). Refers to the portion of the area under a frequency curve which is lying between the start of the distribution and some ordinate lying between the start and the mode ; or symmetrically, between some ordinate lying between the mode and the end of the variate range and the end of the distribution. The term is applicable only to distributions which 'tail off' at their extremes, *i.e.*, have frequency functions tending to zero.

Tally. Refers to a method of keeping count in blocks of five.

1	= 1
11	= 2
111	= 3
1111	= 4
	= 5
1	= 6
11	=12

Tantiles (Tantili). Refer to the $n-1$ partition values of a variable which divide the total amount of an extensive magnitude into a given number n of equal amounts. If $n=2$, the central tantile is termed as dividing value (valore divisorio).

Target Population. Refers to the population about which we wish to get information. Every population must be defined in terms of (1) content, (2) units, (3) extent and (4) time.

Target Sample. This term is used to denote the sample originally selected or designated. Non-response may make the achieved sample to fall short of the target sample.

Telescoping. Refers to a type of memory or recall error. When respondents have been asked to provide information about the past, the question usually involves a reference period—for example, 'How much money did you spend on household repairs in the last three months ?' Frequently the respondent may include in his answer, by mistake, items of expenditure which occurred more than three months previously, particularly if the expenditure had been large. This process is termed as telescoping.

T-Distribution. Refers to the distribution of the means of samples of given size, standardized using estimated values of the parent standard deviation. It has been used in the t-test.

Temporally Continuous Process. Refers to an expression, which is sometimes used to denote a stochastic process which depends upon a continuous time parameter. The terminology is not recommended.

Temporally Homogeneous Process. Refers to a stochastic process for which the transition probabilities have been same for any time interval of length *t*.

Terminal Decision. In sampling schemes of a sequential type, this terms refers to a decisions which involves terminating the sampling process.

Test Coefficient. In factor analysis, refers to a synonyms for factor loading.

Test of Independence. A statistical test of proportions or frequencies, in determine if membership in categories of one variable is different as a function of membership in the categories of a second variable.

Test of Normality. Refers to a test of a set of observations to see whether they could have arisen by random sampling from a Normal population. Such tests may be done by a comparison of the sample distribution function with a Normal distribution function.

Test Statistic. 1. A statistic which has been calculated from a sample in order to test a hypothesis about the parent population from which the sample is drawn ; examples are χ^2, and *t*. 2. . A statistic calculated from more than one sample (like F, U, etc.) to compare the parent populations from which they are drawn.

Theoretical Frequencies. Refers to the frequencies which would fall into assigned ranges of the variate if some theoretical distribution law had been exactly followed, as distinct from the actual frequencies which may be observed in a sample.

Theoretical Variable. This variable is used to denote a variable or variate introduced into structural relations but not directly observable ; as distinct from a 'true' variable or variate, which has been observable except for errors of observation and an 'observable' variable or variate which could be directly observed.

Theory of Runs. A theory developed to allow us to test samples for the randomness of their order.

The Expected Value of Perfect Information. The difference between expected profit (under conditions of risk) and expected profit with perfect information.

Three Dimensional Lattice. Refers to a general class of lattice design of which the cubic lattice has been a particular case. For example, 120 treatments could be laid out as a $4 \times 5 \times 6$ three-dimensional lattice. This would be considered as equivalent to a factorial design having three factors at 4, 5 and 6 levels.

Three Point Assay. Refers to the simplest case of the general class of $(2k+1)$ point slope ralio assays. This design has been useful only where the validity of the assay has been known *a priori*, since in the analysis of the results no degree of freedom have been available for tests of validity.

Three-quarters High Rule. Refers to an empirical rule for ensuring that a bar chart provides the right impression. It states that the maximum height should be three-quarters of the horizontal displacement.

Three-Stage Least Squares. Refers to a method for estimating the coefficients in a simultaneous equations model as a whole than as a set of single equations in isolation.

Three-Stage Sample. Refers to a special case of multistage sampling. The population is assumed to be classified into first-stage units (primary sampling units). Thus the full procedure would be : (1) to select a sample of constituencies—the first stage ; (2) to select a sample of polling districts within each selected constituency—the second stage ; (3) to select a sample of individuals within each selected polling district—the third stage. The method can be applied to any population in which suitable units are available. In a sample of industrial workers, for first stage units might be firms ; the second stage units departments within firms ; and the third stage units the individual workers.

Ticket Sampling. Refers to a method of selecting a random sample in which the characteristics of each member of the population have been noted on a separate card and the required number of cards are drawn from the resulting pack. Also termed as Lottery Sampling.

Tied Ranks. In ranking procedures, two or more objects or persons may achieve levels of sussess (or failure) which are indistinguishable from each other. In this case, it is usual, when calculating rank correlation coefficients, to allot ranks to each equally successful constituent of the group equivalent to the mean of the ranks which they would have received if they were not equally successful.

Tilling. A technique which is used in confluence or bunch map analysis for setting out systematically all the elementary regressions which are found in all possible subsets of variables in a regression equation.

Time Antithesis. Refers to an index number formula which was derived from another formula by interchanging the sub-scripts denoting the base period and the given period and then taking the reciprocal.

Time Sampling. Refers to the sampling of observations of a variate over a period of time, as distinct from the extraction of a sample from a population at a given point of time. Contrast, for example, the sampling of temperature measurements at a particular place on each day for a number of years (time sampling) with the sampling of weights of children at a school on a given day.

Time Sequence. Refers to a set or array of variate values which have been ordered chronologically (*i.e.*, in time order) as distinct from a time series, where a variate has been observed consistently at fixed intervals of time.

Time Series. Refers to a set of values of a variable recorded over a period of time. Examples are the height of Hassan bin Ibrahim measured every 1 January; the cost of a 25 kg sack of potatoes on the first day of every month at a particular market; the mean daily rainfall at a particular weather station taken over every month.

Time-Series Analysis. Not a distinct, single technique, but a set of techniques, the choice of which depends on the periodicity of the series, the number of observations and the purpose of the analysis. The techniques include analysis by moving averages by statistical progression by seasonal variation, and by fitting a trend line.

Time Trend. The trend of a time series.

Tolerance. A term used in statistical quality control to mean the variation from a given norm or standard which is acceptable for the purpose of production, the standard or norm being expressed as the mean or the group. Thus tolerance limits, for example, are those limits within which an article's size, or other measure of component input or output must lie if the article is to be acceptable.

Tolerance Distribution. Refers to the distribution among a number of individuals of the critical level of intensity at which a stimulus will just produce a reaction in each individual.

Although the distribution of these tolerances can be skew, it has been often possible to make it approximately Normal by a simple transformation like the logarithmic.

Tolerance Factor. In quality control, this term refers to the difference between the upper and lower tolerance limits divided by some measure of the variability of the product, usually the standard deviation. Sometimes one half of this quantity may be considered as the tolerance factor, especially where the distribution of the variate under measurement has been symmetrical.

Tolerance Limits. In quality control, the limiting values between which measurements must lie if an article has to be acceptable, as distinct from confidence limits.

Tolerance Number of Defects. Refers to an expression which in better English would be 'tolerable number of defects'. It would be obtained by multiplying the lot tolerance per cent or fraction defective by the size of the lot or batch submitted for inspection.

Total Correlation. The zero order correlation between two variates, *i.e.*, the correlation between the original data rather than between residuals after some common variation has been abstracted.

Total Determination, Coefficient of. In regression analysis, it refers to the square of the coefficients of multiple correlation : R^2. It represents the proportion of the total variance of the dependent variate which has been accounted for by the variation of the independent variables in the multiple correlation.

Total Effect. The total effect of a variable on another in a path model has been the sum of the direct effect and indirect effect. Thus it represents all the influence one variable has on another whether it operates directly or through its impact on some intermediate variable in the model.

Total Regression. Refers to a regression coefficient of zero order, *i.e.*, a coefficient which is involving only one dependent and one independent variate. The expression has been probably better avoided.

Traffic Intensity. Refers to a critical ratio which has been of importance in the analysis of congestion problems ; it is generally given as

$$\rho = \frac{\text{mean rate of arrival (at a queue)}}{\text{mean rate of service}}$$

but may also be stated as the mean service time divided by the mean inter-arrival time.

Tranche. A share, literally a slice.

Transfer Function. Alternatively termed as the frequency response function.

Transformations. Mathematical manipulations for converting one variable into a different form, so we can fit curves as well as lines by regression.

Transformation Set of Latin Squares. If the rows, columns and letters of a Latin Square have been permuted, the resulting set of Latin squares has been termed as the transformation set. In the case of squares of certain sizes, *e.g.*, 6×6 squares, not all the squares of a transformation set would be different.

Transient State. Alternative name for non-recurrent state.

Transition Matrix or Stochastic Matrix. Refers to a matrix relating the probabilities of a stochastic process being in a particular state given its previous state. Symbol : T. It has been used to model a stochastic process. A transition matrix is also associated with every markov chain.

Transition Probability. In the theory of stoachastic processes, the transition probability may be defined as the conditional probability, that a system in state E_i would be in state E_k at some designated later time.

Translation Parameter. Refers to a parameter of location.

T-Ratio Distribution. Refers to the distribution of the ratio of two random variables which follow the bivariate *t*-distribution. It was developed by Press (1969). The distribution arises in Bayesian approaches to inference in econometrics. It can be shown that the distribution is having no moments.

Treatment. In experimentation, this refers to a stimulus which is applied so as to observe the effect on the experimental situation, or to compare its effect with those of other treatments. In practice, 'treatment' may refer to a physical substance, a procedure or anything which has been capable of controlled application according to the requirements of the experiment.

Treatment Mean Square. Refers to a mean square in a variance analysis which is assignable to differences among the effects of one of the experimental treatments.

Tree Diagram. Refers to a diagram in which events or relationships have been represented by points joined by lines. It is laid out so that the rules, 'multiply along the branches' and 'add vertically' apply.

Trend. Refers to the underlying direction and rate of change in a time series, when allowance is made for the effects of seasonal components, cycle components, and irregular variation. The equation of a trend may be termed as the *trend line.* : it need not be straight although it often is.

Trend Empirical. A trend found in a set of data for which the investigator has no logical or theoretical explanation. Such a trend is usually identified simply to eliminate its influence from consideration in the analysis of cyclical movement. See trend, rational.

Trend Fitting. Refers to the general process of representing the trend component of a time series. A trend could be represented by a particular curve form *e.g.*, the logistic, or by a particular form of the general class of polynomial in time, or by a moving average.

Trend Free Series. This is a series in which no trend is present either because the variable concerned historically has fluctuated about a constant level, or because whatever trend was originally in the data has been removed by statistical adjustment.

Trend Linear. A trend in which the rate of change remains constant. If it were plotted on a graph, it would from a straight line.

Trial. In probability theory in 'trial' has been a deliberate attempt to generate an event which is supposed to be happening under a probabilitic scheme ; *e.g*,, the tossing of a coin is a 'trial', the outcome being one of two possible events, a head or a tail.

Triangular Distribution. Refers to a frequency distribution which, when graphed as a frequency polygon, is having a triangular shape. It may be put in the form

$$dF=\frac{2}{a(a+b)}(a+x)dx, \quad -a\leqslant x\leqslant 0\ ;$$

$$=\frac{2}{b(a+b)}(b-x)dx, \quad 0\leqslant x\leqslant b.$$

Triangular Multiply Linked Block Design. For a triangular design to be multiply linked, an extension of singly and doubly linked,

either $r=2\lambda_1-\theta_2$ with $b=n$

or $r=(n-3)\lambda_2-(n-4)\lambda_1$ and $b=\frac{1}{2}(n-1)\times(n-2)$ with the letters denoting replicates (r), blocks (b) and number of treatments (n) in a group.

Triangular (Singly or Doubly) Linked Blocks. Refers to a type of triangular designs.

Trimmed Mean. Refers to a method of reducing the sensitivity of the arithmetic mean to extreme observations. It is involving the discarding a proportion of the outer observations on either side and averaging the remainder. If $0\leqslant\alpha\leqslant 0{\cdot}5$ then the trimmed mean $\overline{x}_\alpha$ has been the same as the arithmetic mean when $\alpha=0$ and converges to the median when α converges to $0{\cdot}5$.

Trinomial Distribution. Refers to the multinomial distribution for three exhaustive and mutually exclusive probabilities.

Triple Comparisons. The statistical model for paired comparisons could be extended to the $({}^{t}_{3})$ triple comparisons that may be formed from t treatments. The asymptotic relative efficiency of triple comparisons, compared with paired comparisons, has been 1·5 and $9/(4\pi)$ when compared with ordinary analysis of variance.

Trough. Refers to an observation in a discontinuous time series which is lower than each of the two neighbouring observations; or, in the continuous case, a point where the series is having a minimum.

True Mean. An alternative, although little used, term for the mean of a population, as distinct from the mean of a sample.

True Regression. Refers to an expression which has been sometimes used in reference to a sample to denote the regression of one variate upon another which would have been obtained if there had been no errors of observation in the independent variable. The usage has been open to misunderstanding owing to possible confusion with the 'parent' regression, as distinct from sample regression, and is probably better avoided.

Truncation. The word finds use in two senses :

1. The omission from a population of all values above or below a fixed value. For example, if the majority of values in a population are below 100 and there are reasonable grounds for believing that values over 100 are not important, all values over 100 may be ignored.

2. The omission of a decimal fraction from a value (as opposed to rounding). For example, the number 15·98 is rounded to 16 as its nearest integer, but when truncated it is expressed as 15.

T Score. A Z score (standard score) that has been modified by multiplying it by ten and adding fifty in order to eliminate negative, zero, and decimal scores.

T Test. A statistical test used to determine the probability that a statistic (summarizing value for a sample) obtained from sample data is merely a reflection of a chance variation in the sample(s) rather than a measure of a true population parameter. In other words, t, which is based upon the ratio of a statistic to its standard error, is used to determine the level of significance of a statistic of a certain size obtained from a sample of a certain size. The t test is used instead of the critical ratio when the size of the sample is small. The sampling distribution of t, sometimes referred to as Student's distribution or Student's t distribution, does not exactly coincide with a normal frequency distribution curve because, with a small sample, the probability of more extreme deviations is greater than with a large sample. As the size of the sample increases, the distribution of t approaches a normal curve. With very large samples the t distribution becomes a normal probability curve. The calculation of t for any statistic takes into account the size of the sample from which the statistic was obtained.

Turning Point. In an ordered series, this term refers to an observation which has been a peak or a trough. When several contiguous values have been equal, and greater than or less than the neighbouring values, a convention has been required to determine which has been regarded as the turning point, *e.g.*, the middle one may be chosen.

Two-By-Two Frequency Table. A term which is used for the presentation in tabular form of data subject to double dichotomy. If each member of a set of n can bear or not bear an attribute A and an attribute B, such a table might be of the form giving below :

	Bearing A	Not Bearing A	Totals
Bearing B	a	b	$a+b$
Not Bearing B	c	d	$c+d$
Totals	$a+c$	$b+d$	$n=a+b+c+d$

Here, for example, b has been the number of members bearing B but not bearing A.

Two-Factor Theory. See single-factor theory.

Two-Phase Sampling. A sampling procedure in which the final sample is not selected directly from the population but from the first-phase sample, through which information is obtained to improve either the selection or the estimation or both.

Two-Sample Test. Refers to a test on two different, non-matched samples. Compare paired-sample test. This term is often used to mean a test of the null hypothesis that two samples have been drawn from parent populations with the same mean, or the same parent population. It is this test which is the subject of the rest of this entry.

Two Sided Test. An alternative name for double tailed test.

Two-Stage Least Squares. Refers to a limited-information method which can he used to estimate the coefficients of any identifiable equation in isolation from the other equations in a simultaneous equations model.

Two Stage Sample. Refers to a simple case of a multi-stage sample. In this case the population to be sampled has been first classified into primary units, each of which consists of a collection of the basic sampling unit, the secondary unit. A sample of these primary units has been taken, constituting the first stage, and these are then subsampled with respect to their secondary units : this constitutes the second stage.

Two-Stage Sampling. A simple case of multistage sampling. The population is classified into groups of elements—the primary sampling units. A sample of these primary sampling units is selected—the first stage ; the selected primary units are then classified into the basic sampling units—the secondary sampling units ; and a subsample of secondary units is then selected from each—this constitutes the second stage.

Two-Tailed Test. A statistical test used to test a hypothesis when the direction of the difference between samples or relationship between variable is not predicted in the hypothesis. That is, the null hypothesis merely states that there will not be a significant difference between the samples, or relationship between the variables, and the alternative hypothesis does not specify which sample will be further in a given direction (score higher, be more favorable, etc) if there is a difference, or if there is a relationship between the variables whether it will be positive or negative. The term *two-tailed* refers to the fact that there is a region of rejection of the null hypothesis at both ends (or tails) of the sampling Distribution.

Two Way Classification. Refers to the classification of a set of observations according to two criteria of classification as, for example, in a double dichotomy or a correlation table.

Type. A word which is finding in several specialised usages in statistics, *e.g.*, in relation to types of frequency function and types of critical region. In older literature two other usages, both obsolescent, are found :

(*a*) in relation to a value like the mean or the median, which is taken as 'typical' of a frequency distribution ;

(*b*) in relation to the central values of the group of a bivariate frequency table, an array of x-values in a group of y centred at the value y_0 being said to be of type y_0.

Type Bias. Bias which may get imparted to an index number because of using a particular type of average on the constituent series. The actual existence of bias is dependent on the nature of the series to which the averaging is applied and also, of course, on what is meant by 'bias' in the context under discussion.

Type 1 and Type 2 Errors. Refers to those errors which occur when, respectively, a true null hypothesis has been rejected and a false null hypothesis has been accepted.

Type One Counter Model. See counter model, type I.

Type Two Counter Model. See counter model, type II.

Type A Distribution. Refers to a form of compound Poisson distribution proposed by Neyman (1933) for use as a contagious distribution. It has to be distinguished from the Gram-Charlier type A expansion of a frequency function. The frequency of the discontinuous variate x (which takes values 0, 1, 2,..., ∞) at j has been the coefficient of t^j in

$$\exp\{-m_1(1-e^{m_2(t-1)})\}.$$

Other forms have been also in use.

Type B Distribution. Refers to a distribution which is formed by the limit of the sum of a large number of independent Neyman type A distributions wherein the second parameter is having a uniform distribution.

Type C Distribution. Refers to a distribution which is formed in a manner similar to the Neyman type B having a beta distribution for the second parameter.

Type I Distribution. Refers to one of the three main types of the Pearson system of frequency distributions. In general it has been a unimodal distribution having limited range. With the origin at the mode, it was written by Pearson in the form

$$dF=k\left(1+\frac{x}{a_1}\right)^{m1}\left(1-\frac{x}{a_2}\right)^{m2}dx, -a_1\leqslant x\leqslant a_2\,;$$

$$m_1, m_2>-1.$$

where $m_1/a_1=m_2/a_2$. With a suitable choice of origin and scale, it has been equivalent to the beta distribution :

$$dF=\frac{1}{B(p, q)}x^{p-1}(1-x)^{q-1}dx,\ \ 0\leqslant x\leqslant 1\ ;\ p, q>0.$$

Type II Distribution. Refers to a particular case of the type I distribution. It may be written :

$$dF=k\left(1-\frac{x^2}{a^2}\right)^m dx,\ \ -a\leqslant x\leqslant a\ ;\ m\geqslant -1\ ;$$

and has been symmetrical, usually platykurtic in shape and with limited range. An alternative way of expressing the distribution has been

$$y = \frac{1}{B(p, p)} x^{p-1}(1-x)^{p-1}, \; 0 \leqslant x \leqslant 1 \; ; p > 0.$$

In the special case when $m=0$ or $p=1$ it becomes the rectangular distribution.

Type III Distribution. This distribution, in the Pearson system of frequency distributions, is having unlimited range in one direction and it is generally unimodal. It may be written :

$$dF = k\left(1+\frac{x}{a}\right)^{\gamma a} e^{-\gamma x}dx, \; -d \leqslant x - \propto \; ; \gamma > 0, a > 0.$$

or, more simply, by choice of a suitable origin and scale :

$$dF = \frac{1}{\lambda(p)} x^{p-1}e^{-x}dx, \; 0 \leqslant x \leqslant \infty \; ; p > 0.$$

The chi-squared distribution has been of this form.

Type IV Distribution. Refers to one of the three main types of the Pearson system of frequency distributions. Its general shape has been that of a unimodal skew distribution with unlimited range in both directions. It may be put as follows :

$$dF = k\left(1+\frac{x^2}{a^2}\right)^{-m} e^{-\mu \tan^{-1}(x/a)}dx, \; \leqslant \propto \leqslant x \leqslant \propto \; ;$$

$$a > 0, \mu > 0.$$

Type V Distribution. Refers to a unimodal distribution of special type in the Pearson system having origin at the start of the distribution. It is usually written in the form.

$$dF = kx^{-p} e^{-\gamma/x} dx, \; 0 \leqslant x \leqslant \infty \; ; \gamma > 0, p > 1.$$

A transformation of type $y = \gamma/x$ turns it into a type III distribution.

Type VI Distribution. Refers to the third of the three main types in the Pearson system of frequency curves. It has been generally unimodal and skew with unlimited range in one direction and may be written as

$$dF = kx^{-q_1}(x-a)^{q_2}dx, \; a \leqslant x \leqslant \infty \; ; q_1 > q_2 + 1.$$

By the substitution $y = a/x$ this distribution can get reduced to the type I form.

Type VII Distribution. Refers to a unimodal symmetrical distribution of special kind in the Pearson system. It is having unlimited range in both directions and may be put as follows :

$$dF = k\left(1+\frac{x^2}{a^2}\right)^{-m} dx, \; -\infty \leqslant x \leqslant \infty \; ; m > \tfrac{1}{2}.$$

The t-distribution has been a special case of this type.

Type VIII Distribution. Refers to a member of the less important group of Pearson curves which may be written :

$$dF=k\left(1+\frac{x}{a}\right)^{-m}dx,\quad -a\leqslant x\leqslant 0\;;\; 0\leqslant m\leqslant 1.$$

Type IX Distribution. It is a less important member of the Pearson system of distribution curves which may be put as follows :

$$dF=k\left(1+\frac{x}{a}\right)^{m}dx,\quad -a\leqslant x\leqslant 0\;;\; m>-1.$$

Type X Distribution. Refers to a distribution of the Pearsonian system which has been the same as the exponential distribution.

Type XI Distribution. Refers to a J-shaped distribution in the Pearson system which may be written :

$$dF=kx^{-m}dx,\quad b\leqslant x\leqslant \infty\;;\; m>0.$$

The start of the distribution has been at an ordinate $x=b$.

Type XII Distribution. Refers to a special distribution in the Pearson system which is having a twisted J-shape and constitutes a particular case of the type I distribution. The form is as follows :

$$dF=\left[\frac{1+\dfrac{x}{a_1}}{1-\dfrac{x}{a_2}}\right]^{m}dx,\quad -a_1\leqslant x\leqslant a_2\;;\; |\,m\,|>1.$$

Type I Error. Refers to an alternative term for α-error or error of the first kind.

Type II Error, Refers to an alternative term for β-error or error of the second kind.

Type III Error. See error of the third kind.

Type I and II Probabilities. See type I sampling.

Type A Region. In the theory of testing statistical hypotheses, this region refers to a locally unbiassed critical region for testing a simple hypothesis specifying one parameter. Regions of this kind have been obtained by maximising the curvature of the power curve at $\theta=\theta_0$ subject to conditions of local unbiassedness and control of errors of the first kind.

If a type A region does not suffer from the restriction of being merely locally unbiassed, but has been in effect, unbiassed everywhere, then it has been known as a type A_1 region.

Type B Region. Refers to an extension of the concept of the type A region to the case of a composite hypothesis.

Type C Region. Refers to an extension of the concept of the unbiassed critical region of type A which was proposed by Neyman and Pearson to cover a simple hypothesis specifying two parameters. A critical region of this class should be of a given size, unbiassed and of best local power in the neighbourhood of the null values of the parameters, say θ_1^0, θ_2^0. The exact determination of type C regions rests upon knowledge of the errors of the second kind and so as to overcome the general absence of this information.

Type D Region. Refers to an unbiassed critical region proposed by Isaacson (1951) which was used for testing simple hypotheses specifying the values of several parameters. The type D region, which has been a generalisation of the type A region, has been one which maximises the curvature of the power surface subject to conditions of size and unbiassedness.

Type 'E' Region. Refers to a development of the type D region but considered to be of doubtful existence for a wide class of multivariate tests (Giri & Kiefer, 1964).

Type I Sampling. A term which has been sometimes used in Bayesian analysis to denote ordinary sampling from a given population. If this population has been itself the result of sampling from a super-population the term type II sampling has been applied to the selection of the population.

The probabilities that govern the two types of sampling have been termed as type I and type II probabilities respectively. The hierarchy could be extended to types III, IV, etc.

Type II Sampling. See type I sampling.

Type A Series. It was introduced by Charlier to denote the expansion of a continuous frequency function as a series of derivatives of the Normal frequency function. It is also termed as the Gram-Charlier series.

Type B Series. A term introduced by Charlier which has been used to denote the expansion of a frequency function in terms of derivatives of a Poisson distribution.

Type C Series. Refers to an expansion of a frequency function which was proposed by Charlier as an alternative to his type A. The latter can produce negative frequencies in the tails of the distribution and type C purports to remove this anomaly. It has not come into general use.

Typical Period (Periodo Tipo). Refers to the period of time to which, in weighted index numbers, the weights of the individual indices have been referred. It may or may not coincide with the base period.

U

U Curve. In statistics, a frequency distribution curve in which the highest frequencies are at the two extreme ends of the curve and the lowest frequencies are in the middle.

Ultimate Cluster. Refers to the set of elements from a particular primary sampling unit which are included in the sample.

Unadjusted Moment. Refers to a moment of a frequency distribution before any adjustment has been made for the effect of grouping the observations, *e.g.*, before the application of Sheppard's corrections.

Unbiased. A sample, estimation procedure or estimator which is free from bias has been said to be unbiased. Thus no systematic distortion or error is present in the procedure. Such a systematic error should get distinguished from a random error, which may give rise to a distortion on any particular occasion but balances out (cancels out) on the average.

Unbiased Confidence Intervals. A system of confidence intervals which has been selective in the sense of covering the true value of the parameter with the assigned probability $1-\alpha$ and has been covering other values as little as possible has been said to be unbiased if $\Pr\{\delta\epsilon\theta \mid \theta\}=1-\alpha\geqslant \Pr\{\delta\epsilon\theta \mid \theta'\}$ where δ denotes a typical interval: θ denotes the true value ond θ' an alternative. This usage of 'unbiased' has been analogous to that for tests of significance.

Unbiased Error. Refers to an error which may be considered as a member drawn at random from an error population having zero mean. Thus in the long run positive and negative errors tend to cancel out in the sense of having a mean which tends to zero.

Unbiased Estimating Equation. Refers to an equation for the estimation of a parameter in which the terms get unbiased estimators of the corresponding parent values. It does not follow that the estimator of the parameter has been then unbiased itself. For example, if the estimator t of a parameter θ has been given by $a-bt=0$, where a and b have been variates, a and b may be unbiased and hence the equation has been unbiased, but the ratio a/b may still give a biased estimator of θ.

Unbiased Estimator. If an estimator is having an expected value, for all sample sizes, equal to the parameter it is designed to estimate, it is an unbiased estimator. Thus in repeated sampling, the estimator takes on average the true value of the parameter. Therefore although for any particular sample the estimate obtained may not be correct (*i.e.*, may not be equal to the parameter being estimated), the average taken over all possible samples will be equal to the parameter. The estimator has been thus free from systematic error or distortion. Also, see biased estimator.

Unbiasedness. The property of being free from bias.

Unbiased Test. A test of significance may get unbiassd in a manner different from the use of that term in estimation. If w has been a critical region of size α for $H_0 : \theta=\theta_0$ against the simple alternative of $H_1 : \theta=\theta_1$ has been such that its power

$$\Pr\{x \in w \mid \theta_1\} \geqslant \alpha$$

it has been said to give an unbiased test of H_0 against H_1. An important class of unbiased tests have been those which have been also uniformly most powerful.

Unequal Subclasses. See disproportionate subclass numbers.

Unidimensionality. Refers to the capacity to be measured in only one dimension. A univariate frequency array, for example, has been a one-dimension array. Although two axes have been used for the purpose of showing frequency values, in practice a single variate could be depicted along a single axis or by using a single vector of frequencies.

Uniform Distribution. Refers to an alternative term for the rectangular distribution.

Uniformly Better Estimator. In comparing two estimators T_1 and T_2 of a population parameter, if, using a particular criterion, T_1 has been at least as good as T_2 at all possible parameter values, and better at some parameter values, we say that T_1 has been uniformly better than T_2.

Uniformly Most Powerful. For tests with a given probability of type I error (or a test of size) α, a test which is having a power on the alternative hypothesis no less than any other test of size a is called a uniformly most powerful (UMP) test.

Uniform Sampling Fraction. If a sample has been selected from a population which has been grouped into strata, in such a way that the number of units selected from each stratum has been proportional to the total number of units in that stratum, the sample has been said to have been selected with a uniform sampling fraction.

Uniform Scores Test. Refers to a nonparametric test of the equality of two circular populations which was proposed by Wheeler and Watson (1964). For independent random samples of size n_1 and n_2 from the respective populations the test statistic, which has been invariant under rotations, may be written in the form

$$R_1^2 = n_1 + \sum_{i=1}^{n_1} \sum_{j=1}^{n_2} \cos \{2\pi(r_i - r_j)/n\}$$

where r_i denotes the rank order of the ith order statistic of the first sample within the combined sample, and $n = n_1 + n_2$.

Uniformity Trial. Refers to an experiment, or set of trials, in which each experimental unit is receiving exactly the same treatment. The object, *inter alia*, may be to estimate some standard characteristic of that treatment or to investigate some aspects of the experimental technique, *e.g.*, plot size or layout in agricultural trials.

Unimodel. Refers to a frequency distribution which is having only one mode. For example, the normal distribution is having only one mode, which coincides with its mean and median, and asymmetric or skewed distributions may also have single modes which do not coincide with their means and medians.

Unit. Refers to a means of measuring or counting data for a particular purpose, or a homogeneous item in which data are either counted or measured. It may be a unit of measurement, like a kilometre, pound, dollar or hour, or a unit counted for measuring frequency, such as a person, a brick, a share or even an examination mark.

Unit Normal Variate. Refers to a variate which has been Normally distributed with zero mean and unit standard deviation. It has been often written $N(x; 0, 1)$ or $N(0, 1)$.

Unitary Sampling. Refers to sampling in which the ultimate units have been directly chosen, as contrasted with a multistage sampling where primary groups have been first chosen.

Univariate (Univariate Distribution). The term univariate simply means having only one variate (or variable). A univariate distribution has been a single distribution having only one variate, which may be either continuous or discontinuous. For example, the heights of a set of children could be represented in the following table :

Height (cm)	*Frequency*
Under 80	2
80 and under 100	10
100 and under 120	27
120 and under 140	39
140 and under 160	28
160 and under 180	9
180 and over	3

In contrast have been bivariate (two-variate), trivariate (three-variate), etc. distributions.

Unlikelihood Ratio. Refers to a type of nonsequential multiple decision rule which was proposed by Hall (1958) where the sample size has been a minimum subject to upper bounds on the probabilities of making incorrect decisions.

Unreduced Designs. Refers to a particular kind of balanced incomplete block design which is derived by taking all possible sets of *k*-treatments from the full set *t*, each set of *k* forming a block. The design gets completed by permuting the blocks and arranging the treatments randomly.

Unweighted Aggregates Index. Uses all the values considered, where each value is of equal importance.

$$\text{Unweighted aggregates quantity index} \frac{\Sigma Q_1}{\Sigma Q_0} \times 100 .$$

Unweighted Average of Relatives Method. To construct an index number, this method first finds the ratio of the current price to the base price for each product, adds the resulting percentage relatives, and then divides by the number of products.

$$\text{Unweighted average of relatives price index} = \frac{\Sigma\left(\frac{P_1}{P_0} \times 100\right)}{n}$$

Unweighted Mean. Refers to a mean of a set of observations in which no weights have been attached to them, except in the trivial sense that each has weight unity.

Unweighted Means Method (in Variance Analysis). In variance analysis, this term refers to a simple method for the analysis of a set of results for which the sub-group frequencies have been unequal. It involves taking the mean values for each subclass and carrying out an ordinary variance analysis on those means.

U-Shaped Distribution. A frequency distribution shaped more or less like a letter U, though not necessarily symmetrical *i.e.* with the maximum frequencies at the two extremes of the range of the variate.

'U' Statistics. Refers to a class of statistic which is occurring in the construction of rank order tests for the two sample problem : Wilcoxon (1945) and Mann and Whitney (1947). They have been of the form

$$U = \sum_{\lambda=1}^{n_1} \sum_{\eta=1}^{n_2} z_{\lambda\eta}$$

where $z=1$ if $x_\lambda < y_\eta$

$=0 \quad x_\lambda > y_\eta$

and $(x_1, \ldots, x_{n1}), (y_1, \ldots y_{n2})$ have been ordered observations from the two samples.

Upper Quartile. See quartile.

Upper-Tailed Test. A one-tailed hypothesis test in which a sample value significantly above the hypothesized population value will lead us to reject the null hypothesis.

Uspensky's Inequality. Refers to an inequality which was developed from the Bienayme-Tchebychev inequality for use in cases where large positive deviations from the mean have been important. It may be written

$$\Pr(x - \bar{x} \leqslant t) - \sigma^2/(\sigma^2+t^2) \qquad \text{if } t<0$$

$$\geqslant 1-[\sigma^2/(\sigma^2+t^2)] \qquad \text{if } t\geqslant 0.$$

This can also be applied to the sum of random variables.

V

Vague Prior. See informative prior.

Validity. 1. In statistics, correspondence between what a measuring device is supposed to measure and what is really measures.

Validation. A procedure which uses external sources independent of the investigation in order to ascertain whether the measurement process is free from bias. Checking the responses obtained in an interview against official records is an example of a validation exercise. The term is also used to denote a check carried out to find out whether a sample is representative of the population from which it was drawn.

Value Added. Net output.

Value Index. Refers to an index number which is formed from the ratio of aggregate values in the given period to the aggregate values in the base period. Strictly speaking it is not an index number as ordinary understood but a value relative.

Value, Mathematical. A quantity or magnitude, represented by a number.

Value, Ultimate. A value or ideal that a society or subgroup in a society regards as fundamental, unchallengeable, and unchanging.

V'anity' Effect. Refers to a form of bias encountered in surveys of human populations which can be introduced into the results through distorted responses by the individuals being questioned. The person questioned fails to provide an accurate reply but makes instead an assertion more pleasing to his personal vanity.

Variable, Discrete. A variable that is composed of a series of distinct and separate units or categories. All qualitative variables are discrete variables, and those quantitative variables consisting of values that are indivisible units (such as individuals) are also discrete variables. There are no values possible in between the units of a discrete variable. The magnitude of a discrete quantitative variable increases by definite steps.

Variable Error. Refers to an error which is occurring in certain types of experiments on living subjects due to factors which vary during the course of an experiment, such as tiredness, or experience.

Variance. A measure of dispersion of statistical values that is the arithmetic mean of the squared deviations from the mean. It is obtained by subtracting each value in a frequency distribution from the arithmetic mean of the distribution to obtain the deviations from the mean. Each deviation is squared (thus eliminating positive and negative signs). The arithmetic mean of these squared deviations is the variance. Thus the variance is the standard deviation squared. It is usually symbolized by σ^2.

Variance, Analysis of. A method of dividing the total variation of observations into components which can be attributed to or associated with particular sources of variation, *e.g.*, the difference between groups or classes used in classifying the observations.

Variance Component. One of the objects of variance analysis have been to split up the sum of squares of observations about their mean into portions which could be assigned to variation between the classes or sub-classes according to which the data have been classified. If the variables defining the classes have been 'fixed', that is to say, if all the classes under consideration actually appear, these constituent parts of the sums of squares indicate through mean-squares the magnitude of class differences, and the extent to which they differ from the residual mean square provides a test of the hypothesis that such differences have been governing the situation.

Variance Ratio Test. Refers to a test which is based on the ratio of two independent statistics, each of which gets distributed as the variance in samples from Normal populations with the same parent variance. Usually the statistics themselves have been quadratic estimators of the parent variance. The test is widely used in variance analysis to test the homogeneity of a set of means.

Variate Difference Method. Refers to a method of analysis of time series which is having a systematic and a random component. It is based essentially on the consideration that if the systematic part of a series could be represented by a polynomial, then successive differencing will eliminate this element and hence allow of the isolation of the random element, or at least the estimation of its variance.

Variate Transformation. Refers to the transformation of one variate into another, generally by a mathematical equation connecting them. The object has been, as a rule, to transform the distribution function of one variate exactly or approximately into a distribution function of known form and properties.

Variation. A measure of dispersion of statistical values that is the sum of the squared deviations from the mean. To obtain the variation, each value in a frequency distribution is subtracted from the arithmetic mean. Each of the resulting deviations is squared (thus eliminating the plus and minus signs). The sum of these squared deviations is the variation. Variation is not used as frequently as variance or standard deviation.

Vector Correlation Coefficient. Refers to a generalisation of the product moment correlation between two variates for the purpose of measuring the relation between a p-way vector variate and a q-way vector variate. If the dispersion matrices of the vectors have been respectively v_1 and v_2 and the covariance matrix of one set with the other is v_3 the vector correlation coefficient has been defined as $=\{ | v_3 | / | v_1 | \; | v_2 | \}^{1/2}$.

Similarly, if the covariance matrix of all $(p+q)$ variates together has been v, the vector alienation coefficient has been $\{ | v | / | v_1 | v_2 | \}^{1/2}$.

Vicious Circle. Intercorrelation between two factors that tends to cause an undesirable growth in the strength of the factors.

Volume Index. Quantity index.

Von Mises Distribution.. Refers to the circular Normal distribution which was derived by von Mises (1918).

Von Neumann's Ratio. The ratio of the mean square successive difference to the variance of a series had been proposed by von Neumann (1914) as a statistic for testing the independence of successive observations in an ordered series for which the underlying distributions is Normal. In large samples from a random series the distribution of the ratio tends to be Normal with mean 2 and variance $4(n-2)/(n^2-1)$ where n denotes the number of observations. The use of the ratio to test independence in a series of observations has been equivalent to the use of the older Abbe-Helmert criterion.

VN Test. Refers to a goodness-of-fit test adapted from the Kolmogorov-Smirnov test by Kuiper (1960) using the statistic

$$V_N = \sup_{-\infty < x < \infty} [F_N(x) - F(x)] - \inf_{-\infty < x < \infty} [F_N(x) - F(x)].$$

It was extended to cover the ordinary two-sample case by Maag & Stephens (1968).

W

WAGR Test. It is a sequential test procedure for $H_0 : p_0 + p_1$ where the proportion has been derived from observations of a Normally distributed variate greater than a given value. The designation has been derived from the names Wald, Arnold, Goldberg and Rushton.

Waiting Line. It is the American equivalent of the English meaning of the English noun 'queue'.

Waiting Time. In the theory of queues this term has been self explanatory, but, in renewal theory residual waiting time has been the period from a given point in time (t) to the next renewal point. Spent waiting time has been the time elapsed since the last renewal. The sum of the two waiting times has been the length of recurrence interval. In the theory of random walks residual waiting time has been termed 'point of first entry' or 'hitting point' for the interval t, ∞.

Wald-Wolfowitz Runs Test. It is a nonparametric test of the identity of the distribution functions of two continuous populations against general alternative hypotheses, which was proposed by Wald & Wolfowitz (1940). A sample of m independent values from the first population gets combined with a sample of n independent values from the second population, the combined sample gets ranked, and the test statistic is computed as the number of runs, where a run may be defined to be a succession of adjacently-ranked values from the same population which are followed and preceded in rank by values from the other population or no values at all.

Wald-Wolfowitz Test. Refers to a large-sample distribution free test of randomness which is based upon serial covariance proposed by Wald and Wolfowitz (1943). For a series of observations $x_1, \ldots, x_n$ measured about their mean the test statistic has been

$$R_k = \sum_{t=1}^{n} x_t x_{t+k},$$

with $x_{n+j} = x_j$, *i.e.*, the formula has been circular.

Walker Probability Function. A function which was derived by Sir Gilbert Walker (1914) in connection with tests of significance for the ordinates of a periodogram. Developing a result due to Schuster (1898) Walker stated that the probability that one value of the intensity (S^2), of the $m=\frac{1}{2}n$ independent values in a Fourier sequence of intensities, shall not get exceeded $4\sigma^2 k/n$ is $1-(1-e^{-k})^m$. The value of this function for arguments k and m has been known as the Walker (Probability) Function.

Waring Distribution. See factorial distribution.

Watson's 'U' Statistic. A statistic which was proposed by Watson for tests of the hypothesis that two independent random samples (m, n) come from the same unknown population. It could be written $U^2=mn(m+n)^{-2}\Sigma(d_i-\bar{d})^2$ where d_i denotes the difference between the pooled sample distribution functions at the ith point. Hence $d_i=n_i/n-m_i/m$ and $\bar{d}=\Sigma d_i/m+n$).

Watson-Williams' Test. In its single-sample form, it is a test of the hypothesis that the mean direction of a Fisher distribution has been equal to a specified direction against the alternative that it has been not. In its two-sample form a test of the hypothesis of equality of two Fisher population mean directions against the alternative of unequal mean directions. The tests had been proposed by Watson & Williams (1956).

Watson's U_N^2 Test. Either of two nonparametric tests proposed by Watson (1961, 1962). For a single random sample of n observations, with order statistics $x_{(1)}, x_{(2)}, \ldots x_{(n)}$, the test of the hypothesis that a population is having a specific distribution function $F_0(x)$ is having a test statistic that may be written in the form

$$U^2=\sum_{i=1}^{n}[u_i-\bar{u}-\{(2i-1)/(2n)\}+\tfrac{1}{2}]^2+1/(12n)$$

where the u_i are the uniform order statistics defined by

$$u_i=F_0(x_{(i)}),\ i=1, 2, \ldots n \text{ and } \bar{u}=\sum_{i=1}^{n}u_i/n.$$

Wedge Plans. Another name for closed sequential t-tests.

Weibull Distribution. The distribution with probability density function :

$$f(x)=\alpha\beta x^{\beta-1}e^{-\alpha x\beta}$$

($x>0$) where α, β are parameters of the distribution. If $\beta=1$, this is the exponential distribution.

Weight. Refers to a number by which a value is multiplied so as to give it a required measure of importance in relationship to other values. The term is used in simple statistics in the calculation of index numbers and averages and in sampling.

Weighted Aggregates Index. Using all the values considered, this index assigns weights to these values.

$$\text{Weighted aggregates price index} = \frac{\Sigma P_1 Q}{\Sigma P_0 Q} X100$$

For a weighted aggregates price index using quantities as weights, obtain the weighted sum of the current year prices by multiplying each price in the index by its associated quantity and summing the results. Then divide this weighted sum by the weighted sum of the base year prices, and multiply the result by 100.

Weighted Average. An average where each item is multiplied by a figure called a weight relating to the item itself in order that a representative average may be obtained. An extreme example is the average fare paid by passengers on a bus service.

Weighted Average of Relatives Method. To construct an index number, this method weights by importance and uses the value of each element in the composite.

Weighted average of relatives price index

$$= \frac{\sum\left[\left(\frac{P_1}{P_0} \times 100\right)(P_n Q_n)\right]}{\Sigma P_n Q_n}$$

With this index, we weight the relative prices by the values for a fixed reference period and divide the weighted sum of relative prices by the sum of the weights. If we use the base year values as weights, we get :

Weighted average of relatlves quantity index

$$= \frac{\sum\left[\left(\frac{Q_1}{Q_0} \times 100\right)(Q_n P_n)\right]}{\Sigma Q_n P_n}$$

In this quantity index, we weight the relative quantities by the values for a fixed reference period and divide the weighted sum by the sum of the weights.

Weighted Battery. Refers to a group of educational or psychological tests wherein the relative importance of each test has been determined by attaching a weight to the score obtained in that test.

Weight Bias. Bias, usually in an index number, due to the use of incorrect or undesirable weights. As the true value of the complete quantity which an index purports to measure is not in general capable of direct measurement, bias in this sense has been to some extent an arbitrary quantity.

Weight Function. A non-negative function which has been used for weighting purposes ; especially in the theory of decision functions, where the word has been often used synonymously with 'loss function'.

Weighted Index Number. Refers to an index number in which the component items have been weighted according to some system of weights reflecting their relative importance. In one sense nearly all index numbers have been weighted by implication ; for example, an index number of prices amalgamates prices per unit of quantity and the size of these units may vary from one commodity to another in such a way as to constitute weighting. It has been however, usual to describe an index as 'weighted' only when weighting coefficients enter explicitly into its definition and calculation.

Weighted Mean. Refers to a result produced by a technique designed to give recognition to the importance of certain factors when compiling the average of a group of values. In general the weighted mean is :

$$\frac{w_1x_1+w_2x_2+\ldots+w_nx_n}{w_1+w_2+\ldots+w_n}$$

where $w_1, w_2, \ldots, w_n$ are the weightings given to the variables $x_1, x_2, \ldots, x_n$.

Weighting Coefficient. The coefficient which is attached to an observation as its weight in a procedure involving weighting.

White Noise. In analogy with the continuous energy distribution in white light from an incandescent body, a covariance stationary stoachastic process which is having equal power in all frequency intervals over a wide frequency range is termed as white noise. The sequence $(X(t), t \geqslant 0)$ is said to be a white noise process if it possesses a constant spectral density function.

Whittle Distribution. Refers to the distribution of the numbers of transitions (p_{ij}) from state i to state j among n consecutive transitions in a Markov chain having N states and transition matrix $P=(p_{ij})$.

Wide Sense Stationary. See covariance stationary process.

Wilks' Criterion. Refers to a criterion of general use in multivariate analysis for testing hypotheses concerning multi-variate Normal populations, specially hypotheses of homogeneity in means or dispersions. The criterion essentially is dependent on the ratio of the determinants of two matrices of sums of squares and products ; the numerator corresponding to a sum-with-classes and the denominator a total sum. It occurs in various form.

Wilks-Lawley U_l Statistic. This is used in multivariate analysis of variance for testing certain hypotheses.

It may be defined as $U_l = \prod_{i=1}^{l} \theta_i$ where θ_i $(i=1, 2, \ldots\ldots, l)$ are the roots of the determinantal equation $|A-\theta(A+C)| = 0$ and l the number of independent gamma variables. A and C have been independent sum of product matrices based on sample observations with n_1 and n_2 degrees of freedom respectively and $l = \min(p, n_1)$. This statistic can be used for testing (*a*) equality of two dispersion matrices, (*b*) equality of the *p*-dimensional mean vectors and (*c*) the independence between the *p*-set and a *q*-set of variates.

Window. See spectral weight function.

Winsorised Estimation. Refers to a method of estimating the mean of a relatively small sample of observations by using linear systematic statistics and replacing extreme observations by those next in magnitude. This process gets associated with the name of C.P. Winsor who gave this approach and it can apply to symmetric or non-symmetric censoring and adjustment.

Within-Group Sum of Squares. In analysis of variance the total sum of squares of the observations from the overall mean can be broken down into two components : (1) the between groups sum of squares and (2) the 'within group sum of squares'. The latter has been the total of the sums of squares of the observations in each group about the group mean.

Within-Group Variance. See intraclass variance.

Wold's Decomposition Theorem. Refers to a theorem which asserts that any univariate non-deterministic stochastic process can get decomposed into a one sided moving average process and a deterministic process.

Working Mean. Refers to an alternative term for arbitrary origin.

Wrapped Distribution. Refers to the name for a distribution on the real line that gets wrapped around the circle of unit radius. If x has been a random variable on the line with distribution function $F(x)$, the random variable x_w of the corresponding wrapped distribution has been given by

$$x_w = x \pmod{2\pi}$$

and the distribution function of x_w has been given by

$$F_w(\theta) = \sum_{k=-\infty}^{\infty} (F(\theta+2\pi k) - F(2\pi k))$$

Wrapped Normal Distribution. Refers to the probability distribution resulting from wrapping a Normal distribution on the real line around the unit circle. If the Normal distribution is having mean zero and variance σ^2, the corresponding wrapped Normal distribution is having a *p.d.f.* given by

$$f_w(\theta)=\frac{1}{\sigma\sqrt{2\pi}}\sum_{k=-\infty}\exp\left(-\frac{1}{2}\frac{(\theta+2\pi k)^2}{\sigma^2}\right),$$

$$0<\theta\leqslant 2\pi.$$

Wrapped Poisson Distribution. Refers to the distribution which is defined by

$$Pr(\theta=2\pi r/m)=\sum_{k=0}^{\infty} p(r+km\,;\,\lambda),$$

$$r=0,1,\ldots m-1,$$

where m has been an integer and p has been the Poisson probability mass function

$$p(x\,;\,\lambda)=e^{-\lambda}\,\lambda^x/x!,\ x=0,1,\ldots;\ \lambda>0.$$

'W' Statistic. The test statistic for a rank sum two sample test on dispersions proposed by Ansari & Bradley (1960). It has been generally written as $W=\Sigma R(Z)$ which has been, effectively, the sum of ranks associated with one sample of observations after the combined samples get ranked.

$W_n{}^2$ Test. See Cramer-von Mises test.

X

X-axis. The horizontal axis of a diagram.

Y

Yates' Correction. Refers to a continuity correction which may be applied when using the chi-squared test on data from a discrete distribution. It has been generally used only in tests on 2×2 contingency tables with small frequencies. It is applied by subtracting 0.5 from the values of O which are greater than E and adding 0 5 to the values of O which are less than E. (O and E are the observed and Expected frequency respectively in the four categories).

Yates-Grundy Estimator. For a sampling design which has the property that the probabilities of any pair of elements occurring together in the sample is positive, unbiased estimators of the variance of an estimated mean or total have been available. For example, the Yates-Grundy estimator of the variance of the total has been as follows.

$$\text{var}(\hat{Y})=\sum_{i=1}^{n}\sum_{j=1}^{n}\frac{(P_iP_j-P_{ij})}{P_{ij}}\left(\frac{Y_i}{P_i}-\frac{Y_j}{P_j}\right)^2.$$

Y-axis. The vertical axis of a diagram.

Y-Intercept. A constant for any given straight line, whose value represents the predicted value of the Y variable when the X variable has a value of 0.

Youden Square. An experimental design which was proposed by Youden in 1937. It is not a square, and would be better known as a 'Youden design'. For example, a design for seven treatments could be put as follows :

A	B	C	D	E	F	G
B	C	D	E	F	G	A
D	E	F	G	A	B	C

which may be considered as three rows of a Latin square—hence the alternative name of incomplete latin square. The above design has been read downwards as comprising seven blocks of three ; in the whole design every treatment and every pair of treatments occur equally often, and the arrangement provides for the analysis of positional effects within blocks.

Yule Distribution. Refers to a distribution which was proposed by Yule (1925) for biological species investigations of the form

$$f(i)=AB(i, \rho+1)$$

where A and ρ has been constants and $B(i, \rho+1)$ has been a beta function. Its use was extended by Simon (1955).

Yule's Equation. A name which is sometimes given to an auto-regressive equation of the second order, *e.g.*,

$$u_t+\alpha u_{t-1}+\beta u_{t-2}=\epsilon_t.$$

Yule-Walker Equations. Refer to a set of p linear relationships between the parameters a_1, a_2, ...a_p of the pth order autoregressive process

$$u_t=a_1u_{t-1}+a_2u_{t-2}+\ldots+a_pu_{t-p}+\epsilon_t$$

in terms of the autocorrelations $\rho_1, \rho_2, \ldots \rho_p$ of the process. The equations had been introduced by Yule (1927) and Walker (1931).

Z

Z-Chart. It is a form of graphic presentation of a time series consisting of three lines which usually take the shape of a letter 'Z'. The lower line has been a plot of original data in the form of a time series; the centre line has been a cumulative total; the upper line consists of a moving total of the original data.

Z-Distribution. Refers to the distribution of a logarithmic transformation of a variance ratio which has been due to R.A. Fisher. If there have been two independent estimates of the population variance s_1^2 and s_2^2 based upon n_1 and n_2 degrees of freedom the function would be defined as

$$z = \frac{1}{2} \log_e \frac{s_1^2}{s_2^2}.$$

Fisher selected the transformation of the variance ratio to z to simplify interpolation in tables of significance points.

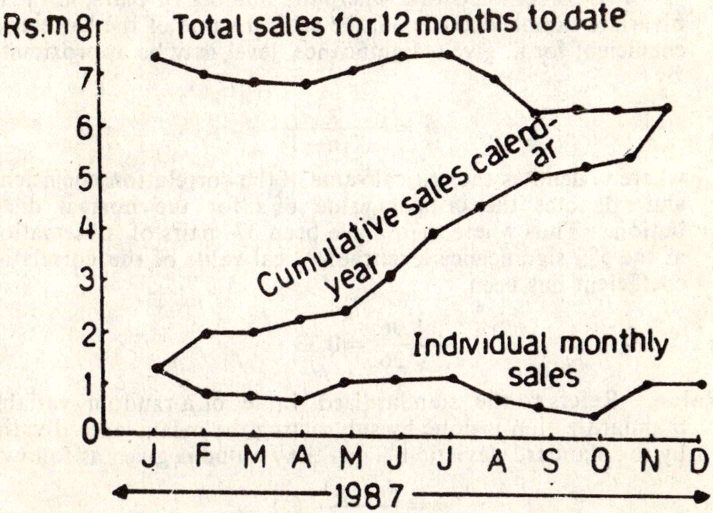

Zeta Distribution. See discrete Pareto distribution.

Zero Sum Game. Refers to a game which is played by a number of persons in which the winner takes all the stakes provided by the losers so that the algebraic sum of gains at any stage is zero. It is argued that many decision problems may be viewed as zero sum games between two persons.

Zipf's Law. A general law which was proposed by Zipf (1949) to approximate to the distribution of individuals bearing variate

Z-Score. A term which is used by some writers in connection with educational and psychological testing as an alternative to standardised scores. A z-score for an observation is the score which is expressed as a deviation from the sample mean value in units of the sample standard deviation.

$$\frac{3.9-9.3}{2.7}=-2$$

Z-score statistics serve to show the extent to which a particular value of a variate differs from the mean of the distribution of the variate.

Z-Transformation of the Correlation Coefficient. A transformation of the correlation coefficient using the formula

$$z=\tanh^{-1}r$$

and taking into account the fact that the distribution of z for samples from a bivariate normal population approaches the normal distribution more quickly than does r.

It is to be noted that where the number of pairs, n, in the bivariate sample exceeds 36, the critical value of the correlation coefficient for a given significance level may be approximated by

$$r_c=\frac{z}{\sqrt{(n-1)}}$$

where r_c denotes the critical value of the correlation coefficient ; and z denotes the critical value of z for the normal distribution. Thus where there have been 37 pairs of observations at the 5% significance level the critical value of the correlation coefficient has been

$$\frac{1.96}{\sqrt{36}}=0.33$$

Z-Value. Refers to the standardized value of a random variable. Standardization is done by subtracting the mean, and dividing by the standard deviation. Thus the z-value is given as follows :

$$z=\frac{(x-\mu)}{\sigma}$$

The z-value is required when using standardized tables such as, for example, those for the normal distribution.